HERBERT SPENCER

ILLUSTRATIONS OF UNIVERSAL PROGRESS

A SERIES OF DISCUSSIONS

Elibron Classics
www.elibron.com

Elibron Classics series.

ISBN 1-4021-9953-8 (paperback)
ISBN 1-4212-9980-1 (hardcover)

This Elibron Classics book contains the complete text of the edition published in 1890 by D. Appleton and Company, New York.

CONTENTS

I
PROGRESS: ITS LAW AND CAUSE

The current conception of Progress is somewhat shifting and indefinite. Sometimes it comprehends little more than simple growth — as of a nation in the number of its members and the extent of territory over which it has spread. Sometimes it has reference to quantity of material products — as when the advance of agriculture and manufactures is the topic. Sometimes the superior quality of these products is contemplated: and sometimes the new or improved appliances by which they are produced. When, again, we speak of moral or intellectual progress, we refer to the state of the individual or people exhibiting it; while, when the progress of Knowledge, of Science, of Art, is commented upon, we have in view certain abstract results of human thought and action. Not only, however, is the current conception of Progress more or less vague, but it is in great measure erroneous. It takes in not so much the reality of Progress as its accompaniments — not so much the substance as the shadow. That progress in intelligence seen during the growth of the child into the man, or the savage into the philosopher, is commonly regarded as consisting in the greater number of facts known and laws understood: whereas the actual progress consists in those internal modifications of which this increased knowledge is the expression. Social progress is supposed to consist in the produce of a greater quantity and variety of the articles required for satisfying men's wants; in the increasing security of person and property; in widening freedom of action: whereas, rightly understood, social progress consists in those changes of structure in the social organism which have entailed these consequences. The current conception is a teleological one. The phenomena are contemplated solely as bearing on human happiness. Only those changes are held to constitute progress which directly or indirectly tend to heighten human happiness. And they are thought to constitute progress simply *because* they tend to heighten human happiness. But rightly to understand progress, we must inquire what is the nature of these changes, considered apart from our interests. Ceasing, for example, to regard the successive geological modifications that have taken place in the Earth, as modifications that have gradually fitted it for the habitation of Man, and as *therefore* a geological progress, we must seek to determine the character common to these modifications — the law to which they all conform. And similarly in every other case. Leaving out of sight concomitants and beneficial consequences, let us ask what Progress is in itself.

In respect to that progress which individual organisms display in the course of their evolution, this question has been answered by the Germans. The investigations of Wolff, Goethe, and Von Baer, have established the truth that the series of changes gone through during the development of a seed into a tree, or an ovum into an animal, constitute an advance from homogeneity of structure to heterogeneity of structure. In its primary stage, every germ consists of a substance that is uniform throughout, both in texture and chemical composition. The first step is the appearance of a difference between two parts of this substance; or, as the phenomenon is called in physiological language, a differentiation. Each of these differentiated divisions presently begins itself to exhibit some contrast of parts; and by and by these secondary differentiations become as definite as the original one. This process is continuously repeated — is simultaneously going on in all parts of the growing embryo; and by endless such differentiations there is finally produced that complex combination of tissues and organs constituting the adult animal or plant. This is the history of all organisms whatever. It is settled beyond dispute that organic progress consists in a change from the homogeneous to the heterogeneous.

Now, we propose in the first place to show, that this law of organic progress is the law of all progress. Whether it be in the development of the Earth, in the development of Life upon its surface, in the development of Society, of Government, of Manufactures, of Commerce, of Language, Literature, Science, Art, this same evolution of the simple into the complex, through successive differentiations, holds throughout. From the earliest traceable cosmical changes down to the latest results of civilization, we shall find that the transformation of the homogeneous into the heterogeneous, is that in which Progress essentially consists.

With the view of showing that *if* the Nebular Hypothesis be true, the genesis of the solar system supplies one illustration of this law, let us assume that the matter of which the sun and planets consist was once in a diffused form; and that from the gravitation of its atoms there resulted a gradual concentration. By the hypothesis, the solar system in its nascent state existed as an indefinitely extended and nearly homogeneous medium — a medium almost homogeneous in density, in temperature, and in other physical attributes. The first advance towards consolidation resulted in a differentiation between the occupied space which the nebulous mass still filled, and the unoccupied space which it previously filled. There simultaneously resulted a contrast in density and a contrast in temperature, between the interior and the exterior of this mass. And at the same time there arose throughout it rotatory movements, whose velocities varied according to their distances from its centre. These differentiations increased in number and degree until there was evolved the organized group of sun, planets, and satellites, which we now know — a group which presents numerous contrasts of structure and action among its members. There are the immense contrasts between the sun and planets, in bulk and in weight; as well as the subordinate contrasts between one planet and another, and between the planets and their satellites. There is the similarly marked contrast between the sun as almost stationary, and the planets as moving round him with great velocity; while there are the secondary contrasts between the velocities and periods of the several planets, and between their simple revolutions and the double ones of their satellites, which have to move round their primaries while moving round the sun. There is the yet further strong contrast between the sun and the planets in respect of temperature; and there is reason to suppose that the planets and satellites differ from each other in their proper heat, as well as in the heat they receive from the sun.

When we bear in mind that, in addition to these various contrasts, the planets and satellites also differ in respect to their distances from each other and their primary; in respect to the inclinations of their orbits, the inclinations of their axes, their times of rotation on their axes, their specific gravities, and their physical constitutions; we see what a high degree of heterogeneity the solar system exhibits, when compared with the almost complete homogeneity of the nebulous mass out of which it is supposed to have originated.

Passing from this hypothetical illustration, which must be taken for what it is worth, without prejudice to the general argument, let us descend to a more certain order of evidence. It is now generally agreed among geologists that the Earth was at first a mass of molten matter; and that it is still fluid and incandescent at the distance of a few miles beneath its surface. Originally, then, it was homogeneous in consistence, and, in virtue of the circulation that takes place in heated fluids, must have been comparatively homogeneous in temperature; and it must have been surrounded by an atmosphere consisting partly of the elements of air and water, and partly of those various other elements which assume a gaseous form at high temperatures. That slow cooling by radiation which is still going on at an inappreciable rate, and which, though originally far more rapid than now, necessarily required an immense time to produce any decided change, must ultimately have resulted in the solidification of the portion most able to part with its heat — namely, the surface. In the thin crust thus formed we have the first marked differentiation. A still further cooling, a consequent thickening of this crust, and an accompanying deposition of all solidifiable elements contained in the atmosphere, must finally have been followed by the condensation of the water previously existing as vapour. A second marked differentiation must thus have arisen: and as the condensation must have taken place on the coolest parts of the surface — namely, about the poles — there must thus have resulted the first geographical distinction of parts. To these illustrations of growing heterogeneity, which, though deduced from the known laws of matter, may be regarded as more or less hypothetical, Geology adds an extensive series that have been inductively established. Its investigations show that the Earth has been continually becoming more heterogeneous in virtue of the multiplication of the strata which form its crust; further, that it has been becoming more heterogeneous in respect of the composition of these strata, the latter of which, being made from the detritus of the older ones, are many of them rendered highly complex by the mixture of materials they contain;

and that this heterogeneity has been vastly increased by the action of the Earth's still molten nucleus upon its envelope, whence have resulted not only a great variety of igneous rocks, but the tilting up of sedimentary strata at all angles, the formation of faults and metallic veins, the production of endless dislocations and irregularities. Yet again, geologists teach us that the Earth's surface has been growing more varied in elevation — that the most ancient mountain systems are the smallest, and the Andes and Himalayas the most modern; while in all probability there have been corresponding changes in the bed of the ocean. As a consequence of these ceaseless differentiations, we now find that no considerable portion of the Earth's exposed surface is like any other portion, either in contour, in geologic structure, or in chemical composition; and that in most parts it changes from mile to mile in all these characteristics.

Moreover, it must not be forgotten that there has been simultaneously going on a gradual differentiation of climates. As fast as the Earth cooled and its crust solidified, there arose appreciable differences in temperature between those parts of its surface most exposed to the sun and those less exposed. Gradually, as the cooling progressed, these differences became more pronounced; until there finally resulted those marked contrasts between regions of perpetual ice and snow, regions where winter and summer alternately reign for periods varying according to the latitude, and regions where summer follows summer with scarcely an appreciable variation. At the same time the successive elevations and subsidences of different portions of the Earth's crust, tending as they have done to the present irregular distribution of land and sea, have entailed various modifications of climate beyond those dependent on latitude; while a yet further series of such modifications have been produced by increasing differences of elevation in the land, which have in sundry places brought arctic, temperate, and tropical climates to within a few miles of each other. And the general result of these changes is, that not only has every extensive region its own meteorologic conditions, but that every locality in each region differs more or less from others in those conditions, as in its structure, its contour, its soil. Thus, between our existing Earth, the phenomena of whose varied crust neither geographers, geologists, mineralogists, nor meteorologists have yet enumerated, and the molten globe out of which it was evolved, the contrast in heterogeneity is sufficiently striking.

When from the Earth itself we turn to the plants and animals that have lived, or still live, upon its surface, we find ourselves in some difficulty from lack of facts. That every existing organism has been developed out of the simple into the complex, is indeed, the first established truth of all; and that every organism that has existed was similarly developed, is an inference which no physiologist will hesitate to draw. But when we pass from individual forms of life to Life in general, and inquire whether the some law is seen in the *ensemble* of its manifestations, — whether modern plants and animals are of more heterogeneous structure than ancient ones, and whether the Earth's present Flora and Fauna are more heterogeneous than the Flora and Fauna of the past, — we find the evidence so fragmentary, that every conclusion is open to dispute. Two-thirds of the Earth's surface being covered by water; a great part of the exposed land being inaccessible to, or untravelled by, the geologist; the greater part of the remainder having been scarcely more than glanced at; and even the most familiar portions, as England, having been so imperfectly explored that a new series of strata has been added within these four years, — it is manifestly impossible for us to say with any certainty what creatures have, and what have not, existed at any particular period. Considering the perishable nature of many of the lower organic forms, the metamorphosis of many sedimentary strata, and the gaps that occur among the rest, we shall see further reason for distrusting our deductions. On the one hand, the repeated discovery of vertebrate remains in strata previously supposed to contain none, — of reptiles where only fish were thought to exist, — of mammals where it was believed there were no creatures higher than reptiles, — renders it daily more manifest how small is the value of negative evidence.

On the other hand, the worthlessness of the assumption that we have discovered the earliest, or anything like the earliest, organic remains, is becoming equally clear. That the oldest known sedimentary rocks have been greatly changed by igneous action, and that still older ones have been totally transformed by it, is becoming undeniable. And the fact that sedimentary strata earlier than any we know, have been melted up, being admitted, it must also be admitted that we cannot say how far back in time this destruction of sedimentary strata has been going on. Thus it is manifest that the title, *Palœozoic*, as applied to the earliest known fossiliferous strata, involves a *petitio principii;* and that, for aught we know to the contrary, only the last few chapters of the Earth's biological history may have come down to us. On neither side, therefore, is the evidence conclusive. Nevertheless we cannot but think that, scanty as they are, the facts, taken altogether, tend to show both that the more heterogeneous organisms have been evolved in the later geologic periods, and that Life in general has been more heterogeneously manifested as time has advanced. Let us cite, in illustration, the one case of the *vertebrata*. The earliest known vertebrate remains are those of Fishes; and Fishes are the most Homogeneous of the vertebrata. Later and more heterogeneous are Reptiles. Later still, and more heterogeneous still, are Mammals and Birds. If it be said, as it may fairly be said, that the Palœozoic deposits, not being estuary deposits, are not likely to contain the remains of terrestrial vertebrata, which may nevertheless have existed at that era, we reply that we are merely pointing to the leading facts, *such as they are*.

But to avoid any such criticism, let us take the mammalian subdivision only. The earliest known remains of mammals are those of small marsupials, which are the lowest of the mammalian type; while, conversely, the highest of the mammalian type — Man — is the most recent. The evidence that the vertebrate fauna, as a whole, has become more heterogeneous, is considerably stronger. To the argument that the vertebrate fauna of the Palœozoic period, consisting, so far as we know, entirely of Fishes, was less heterogeneous than the modern vertebrate fauna, which includes Reptiles, Birds, and Mammals, of multitudinous genera, it may be replied, as before, that estuary deposits of the Palœozoic period, could we find them, might contain other orders of vertebrata. But no such reply can be made to the argument that whereas the marine vertebrata of the Palœozoic period consisted entirely of cartilaginous fishes, the marine vertebrata of later periods include numerous genera of osseous fishes; and that, therefore, the later marine vertebrate faunas are more heterogeneous than the oldest known one. Nor, again, can any such reply be made to the fact that there are far more numerous orders and genera of mammalian remains in the tertiary formations than in the secondary formations. Did we wish merely to make out the best case, we might dwell upon the opinion of Dr. Carpenter, who says that "the general facts of Palæontology appear to sanction the belief, that *the same plan* may be traced out in what may be called *the general life of the globe*, as in the individual life of every one of the forms of organized being which now people it." Or we might quote, as decisive, the judgment of Professor Owen, who holds that the earlier examples of each group of creatures severally departed less widely from archetypal generality than the later ones — were severally less unlike the fundamental form common to the group as a whole; that is to say — constituted a less heterogeneous group of creatures; and who further upholds the doctrine of a biological progression. But in deference to an authority for whom we have the highest respect, who considers that the evidence at present obtained does not justify a verdict either way, we are content to leave the question open.

Whether an advance from the homogeneous to the heterogeneous is or is not displayed in the biological history of the globe, it is clearly enough displayed in the progress of the latest and most heterogeneous creature — Man. It is alike true that, during the period in which the Earth has been peopled, the human organism has grown more heterogeneous among the civilized divisions of the species; and that the species, as a whole, has been growing more heterogeneous in virtue of the multiplication of races and the differentiation of these races from each other.

In proof of the first of these positions, we may cite the fact that, in the relative development of the limbs, the civilized man departs more widely from the general type of the placental mammalia than do the lower human races. While often possessing well-developed body and arms, the Papuan has extremely small legs: thus reminding us of the quadrumana, in which there is no great contrast in size between the hind and fore limbs. But in the European, the greater length and massiveness of the legs has become very marked — the fore and hind limbs are relatively more heterogeneous. Again, the greater ratio which the cranial bones bear to the facial bones illustrates the same truth. Among the vertebrata in general, progress is marked by an increasing heterogeneity in the vertebral column, and more especially in the vertebrae constituting the skull: the higher forms being distinguished by the relatively larger size of the bones which cover the brain, and the relatively smaller size of those which form the jaw, &c. Now, this characteristic, which is stronger in Man than in any other creature, is stronger in the European than in the savage. Moreover, judging from the greater extent and variety of faculty he exhibits, we may infer that the civilized, man has also a more complex or heterogeneous nervous system than the uncivilized man: and indeed the fact is in part visible in the increased ratio which his cerebrum bears to the subjacent ganglia.

If further elucidation be needed, we may find it in every nursery. The infant European has sundry marked points of resemblance to the lower human races; as in the flatness of the also of the nose, the depression of its bridge, the divergence and forward opening of the nostrils, the form of the lips, the absence of a frontal sinus, the width between the eyes, the smallness of the legs. Now, as the developmental process by which these traits are turned into those of the adult European, is a continuation of that change from the homogeneous to the heterogeneous displayed during the previous evolution of the embryo, which every physiologist will admit; it follows that the parallel developmental process by which the like traits of the barbarous races have been turned into those of the civilized races, has also been a continuation of the change from the homogeneous to the heterogeneous. The truth of the second position — that Mankind, as a whole, have become more heterogeneous — is so obvious as scarcely to need illustration. Every work on Ethnology, by its divisions and subdivisions of races, bears testimony to it. Even were we to admit the hypothesis that Mankind originated from several separate stocks, it would still remain true, that as, from each of these stocks, there have sprung many now widely different tribes, which are proved by philological evidence to have had a common origin, the race as a whole is far less homogeneous than it once was. Add to which, that we have, in the Anglo-Americans, an example of a new variety arising within these few generations; and that, if we may trust to the description of observers, we are likely soon to have another such example in Australia.

On passing from Humanity under its individual form, to Humanity as socially embodied, we find the general law still more variously exemplified. The change from the homogeneous to the heterogeneous is displayed equally in the progress of civilization as a whole, and in the progress of every tribe or nation; and is still going on with increasing rapidity. As we see in existing barbarous tribes, society in its first and lowest, form is a homogeneous aggregation of individuals having like powers and like functions: the only marked difference of function being that which accompanies difference of sex. Every man is warrior, hunter, fisherman, tool-maker, builder; every woman performs the same drudgeries; every family is self-sufficing, and save for purposes of aggression and defence, might as well live apart from the rest. Very early, however, in the process of social evolution, we find an incipient differentiation between the governing and the governed. Some kind of chieftainship seems coeval with the first advance from the state of separate wandering families to that of a nomadic tribe. The authority of the strongest makes itself felt among a body of savages as in a herd of animals, or a posse of schoolboys. At first, however, it is indefinite, uncertain; is shared by others of scarcely inferior power: and is unaccompanied by any difference in occupation or style of living: the first ruler kills his own game, makes his own weapons, builds his own hut, and economically considered, does not differ from others of his tribe. Gradually, as the tribe progresses, the contrast between the governing and the governed grows more decided. Supreme power becomes hereditary in one family; the head of that family, ceasing to provide for his own wants, is served by others; and he begins to assume the sole office of ruling.

At the same time there has been arising a co-ordinate species of government — that of Religion. As all ancient records and traditions prove, the earliest rulers are regarded as divine personages. The maxims and commands they uttered during their lives are held sacred after their deaths, and are enforced by their divinely-descended successors; who in their turns are promoted to the pantheon of the race, there to be worshipped and propitiated along with their predecessors: the most ancient of whom is the supreme god, and the rest subordinate gods. For a long time these connate forms of government — civil and religious — continue closely associated. For many generations the king continues to be the chief priest, and the priesthood to be members of the royal race. For many ages religious law continues to contain more or less of civil regulation, and civil law to possess more or less of religious sanction; and even among the most advanced nations these two controlling agencies are by no means completely differentiated from each other.

Having a common root with these, and gradually diverging from them, we find yet another controlling agency — that of Manners or ceremonial usages. All titles of honour are originally the names of the god-king; afterwards of God and the king; still later of persons of high rank; and finally come, some of them, to be used between man and man. All forms of complimentary address were at first the expressions of submission from prisoners to their conqueror, or from subjects to their ruler, either human or divine — expressions that were afterwards used to propitiate subordinate authorities, and slowly descended into ordinary intercourse. All modes of salutation were once obeisances made before the monarch and used in worship of him after his death. Presently others of the god-descended race were similarly saluted; and by degrees some of the salutations have become the due of all.[1] Thus, no sooner does the originally homogeneous social mass differentiate into the governed and the governing parts, than this last exhibits an incipient differentiation into religions and secular — Church and State; while at the same time there begins to be differentiated from both, that less definite species of government which rules our daily intercourse — a species of government which, as we may see in heralds' colleges, in books of the peerage, in masters of ceremonies, is not without a certain embodiment of its own. Each of these is itself subject to successive differentiations. In the course of ages, there arises, as among ourselves, a highly complex political organization of monarch, ministers, lords and commons, with their subordinate administrative departments, courts of justice, revenue offices, &c., supplemented in the provinces by municipal governments, county governments, parish or union governments — all of them more or less elaborated. By its side there grows up a highly complex religious organization, with its various grades of officials, from archbishops down to sextons, its colleges, convocations, ecclesiastical courts, &c.; to all which must be added the ever multiplying independent sects, each with its general and local authorities. And at the same time there is developed a

---

[1] For detailer proof of these assertions see essay on Manners and Fashion.

highly complex aggregation of customs, manners, and temporary fashions, enforced by society at large, and serving to control those minor transactions between man and man which are not regulated by civil and religious law. Moreover it is to be observed that this ever increasing heterogeneity in the governmental appliances of each nation, has been accompanied by an increasing heterogeneity in the governmental appliances of different nations; all of-which are more or less unlike in their political systems and legislation, in their creeds and religious institutions, in their customs and ceremonial usages.

Simultaneously there has been going on a second differentiation of a more familiar kind; that, namely, by which the mass of the community has been segregated into distinct classes and orders of workers. While the governing part has undergone the complex development above detailed, the governed part has undergone an equally complex development, which has resulted in that minute division of labour characterizing advanced nations. It is needless to trace out this progress from its first stages, up through the caste divisions of the East and the incorporated guilds of Europe, to the elaborate producing and distributing organization existing among ourselves. Political economists have long since described the evolution which, beginning with a tribe whose members severally perform the same actions each for himself, ends with a civilized community whose members severally perform different actions for each other; and they have further pointed out the changes through which the solitary producer of any one commodity is transformed into a combination of producers who, united under a master, take separate parts in the manufacture of such commodity. But there are yet other and higher phases of this advance from the homogeneous to the heterogeneous in the industrial organization of society.

Long after considerable progress has been made in the division of labour among different classes of workers, there is still little or no division of labour among the widely separated parts of the community; the nation continues comparatively homogeneous in the respect that in each district the same occupations are pursued. But when roads and other means of transit become numerous and good, the different districts begin to assume different functions, and to become mutually dependent. The calico manufacture locates itself in this county, the woollen-cloth manufacture in that; silks are produced here, lace there; stockings in one place, shoes in another; pottery, hardware, cutlery, come to have their special towns; and ultimately every locality becomes more or less distinguished from the rest by the leading occupation carried on in it. Nay, more, this subdivision of functions shows itself not only among the different parts of the same nation, but among different nations. That exchange of commodities which free-trade promises so greatly to increase, will ultimately have the effect of specializing, in a greater or less degree, the industry of each people. So that beginning with a barbarous tribe, almost if not quite homogeneous in the functions of its members, the progress has been, and still is, towards an economic aggregation of the whole human race; growing ever more heterogeneous in respect of the separate functions assumed by separate nations, the separate functions assumed by the local sections of each nation, the separate functions assumed by the many kinds of makers and traders in each town, and the separate functions assumed by the workers united in producing each commodity.

Not only is the law thus clearly exemplified in the evolution of the social organism, but it is exemplified with equal clearness in the evolution of all products of human thought and action, whether concrete or abstract, real or ideal. Let us take Language as our first illustration.

The lowest form of language is the exclamation, by which an entire idea is vaguely conveyed through a single sound; as among the lower animals. That human language ever consisted solely of exclamations, and so was strictly homogeneous in respect of its parts of speech, we have no evidence. But that language can be traced down to a form in which nouns and verbs are its only elements, is an established fact. In the gradual multiplication of parts of speech cut of these primary ones — in the differentiation of verbs into active and passive, of nouns into abstract and concrete — in the rise of distinctions of mood, tense, person, of number and case — in the formation of auxiliary verbs, of adjectives, adverbs, pronouns, prepositions, articles — in the divergence of those orders, genera, species, and varieties of parts of speech by which civilized races express minute modifications of meaning — we see a change from the homogeneous to the heterogeneous. And it may be remarked, in passing, that it is more especially in virtue of having carried this subdivision of function to a greater extent and completeness, that the English language is superior to all others.

Another aspect under which we may trace the development of language is the differentiation of words of allied meanings. Philology early disclosed the truth that in all languages words may be grouped into families having a common ancestry. An aboriginal name applied indiscriminately to each of an extensive and ill-defined class of things or actions, presently undergoes modifications by which the chief divisions of the class are expressed. These several names springing from the primitive root, themselves become the parents of other names still further modified. And by the aid of those systematic modes which presently arise, of making derivations and forming compound terms expressing still smaller distinction there is finally developed a tribe of words so heterogeneous in sound and meaning, that to the uninitiated it seems incredible that they should have had a common origin. Meanwhile from other roots there are being evolved other such tribes, until there results a language of some sixty thousand or more unlike words, signifying as many unlike objects, qualities, acts.

Yet another way in which language in general advances from the homogeneous to the heterogeneous, is in the multiplication of languages. Whether as Max Müller and Bunsen think, all languages have grown from one stock, or whether, as some philologists say, they have grown from two or more stocks, it is clear that since large families of languages, as the Indo-European, are of one parentage, they have become distinct through a process of continuous divergence. The same diffusion over the Earth's surface which has led to the differentiation of the race, has simultaneously led to a differentiation of their speech: a truth which we see further illustrated in each nation by the peculiarities of dialect found in several districts. Thus the progress of Language conforms to the general law, alike in the evolution of languages, in the evolution of families of words, and in the evolution of parts of speech.

On passing from spoken to written language, we come upon several classes of facts, all having similar implications. Written language is connate with Painting and Sculpture; and at first all three are appendages of Architecture, and have a direct connection with the primary form of all Government — the theocratic.

Merely noting by the way the fact that sundry wild races, as for example the Australians and the tribes of South Africa, are given to depicting personages and events upon the walls of caves, which are probably regarded as sacred places, let us pass to the case of the Egyptians. Among them, as also among the Assyrians, we find mural paintings used to decorate the temple of the god and the palace of the king (which were, indeed, originally identical); and as such they were governmental appliances in the same sense that state-pageants and religious feasts were. Further, they were governmental appliances in virtue of representing the worship of the god, the triumphs of the god-king, the submission of his subjects, and the punishment of the rebellious. And yet again they were governmental, as being the products of an art reverenced by the people as a sacred mystery. From the habitual use of this pictorial representation there naturally grew up the but slightly-modified practice of picture-writing — a practice which was found still extant among the Mexicans at the time they were discovered. By abbreviations analogous to those still going on in our own written and spoken language, the most familiar of these pictured figures were successively simplified; and ultimately there grew up a system of symbols, most of which had but a distant resemblance to the things for which they stood. The inference that the hieroglyphics of the Egyptians were thus produced, is confirmed by the fact that the picture-writing of the Mexicans was found to have given birth to a like family of ideographic forms; and among them, as among the Egyptians, these had been partially differentiated into the *kuriological* or imitative, and the *tropical* or symbolic: which were, however, used together in the same record. In Egypt, written language underwent a further differentiation: whence resulted the *hieratic* and the *epistolographic* or *enchorial*: both of which are derived from the original hieroglyphic. At the same time we find that for the expression of proper names which could not be otherwise conveyed, phonetic symbols were employed; and though it is alleged that the Egyptians never actually achieved complete alphabetic writing, yet it can scarcely be doubted that these phonetic symbols occasionally used in aid of their ideographic ones, were the germs out of which alphabetic writing grew. Once having become separate from hieroglyphics, alphabetic writing itself underwent numerous differentiations — multiplied alphabets were produced; between most of which, however, mere or less connection can still be traced. And in each civilized nation there has now grown up, for the representation of one set of sounds, several sets of written signs used for distinct purposes. Finally, through a yet more important differentiation came printing; which, uniform in kind as it was at first, has since become multiform.

While written language was passing through its earlier stages of development, the mural decoration which formed its root was being differentiated into Painting and Sculpture. The gods, kings, men, and animals represented, were originally marked by indented outlines and coloured. In most cases these outlines were of such depth, and the object they circumscribed so far rounded and marked out in its leading parts, as to form a species of work intermediate between intaglio and bas-relief. In other cases we see an advance upon this: the raised spaces between the figures being chiselled off, and the figures themselves appropriately tinted, a painted bas-relief was produced. The restored Assyrian architecture at Sydenham exhibits this style of art carried to greater perfection — the persons and things represented, though still barbarously coloured, are carved out with more truth and in greater detail: and in the winged lions and bulls used for the angles of gateways, we may see a considerable advance towards a completely sculptured figure; which, nevertheless, is still coloured, and still forms part of the building. But while in Assyria the production of a statue proper seems to have been little, if at all, attempted, we may trace in Egyptian art the gradual separation of the sculptured figure from the wall. A walk through the collection in the British Museum will clearly show this; while it will at the same time afford an opportunity of observing the evident traces which the independent statues bear of their derivation from bas-relief; seeing that nearly all of them not only display that union of the limbs with the body which is the characteristic of bas-relief, but have the back of the statue united from bead to foot with a block which stands in place of the original wall. Greece repeated the leading stages of this progress. As in Egypt and Assyria, these twin arts were at first united with each other and with their parent, Architecture, and were the aids of Religion and Government. On the friezes of Greek temples, we see coloured bas-reliefs representing sacrifices, battles, processions, games — all in some sort religious. On the pediments we see painted sculptures more or less united with the tympanum, and having for subjects the triumphs of gods or heroes. Even when we come to statues that are definitely separated from the buildings to which they pertain, we still find them coloured; and only in the later periods of Greek civilization does the differentiation of sculpture from painting appear to have become complete.

In Christian art we may clearly trace a parallel re-genesis. All early paintings and sculptures throughout Europe were religious in subject — represented Christs, crucifixions, virgins, holy families, apostles, saints. They formed integral parts of church architecture, and were among the means of exciting worship; as in Roman Catholic countries they still are. Moreover, the early sculptures of Christ on the cross, of virgins, of saints, were coloured: and it needs but to call to mind the painted madonnas and crucifixes still abundant in continental churches and highways, to perceive the significant fact that painting and sculpture continue in closest connection with each other where they continue in closest connection with their parent. Even when Christian sculpture was pretty clearly differentiated from painting, it was still religious and governmental in its subjects — was used for tombs in churches and statues of kings: while, at the same time, painting, where not purely ecclesiastical, was applied to the decoration of palaces, and besides representing royal personages, was almost wholly devoted to sacred legends. Only in quite recent times have painting and sculpture become entirely secular arts. Only within these few centuries has painting been divided into historical, landscape, marine, architectural, genre, animal, still-life, &c., and sculpture grown heterogeneous in respect of the variety of real and ideal subjects with which it occupies itself.

Strange as it seems then, we find it no less true, that all forms of written language, of painting, and of sculpture, have a common root in the politico-religious decorations of ancient temples and palaces. Little resemblance as they now have, the bust that stands on the console, the landscape that hangs against the wall, and the copy of the *Times* lying upon the table, are remotely akin; not only in nature, but by extraction. The brazen face of the knocker which the postman has just lifted, is related not only to the woodcuts of the *Illustrated London News* which he is delivering, but to the characters of the *billet-doux* which accompanies it. Between the painted window, the prayer-book on which its light falls, and the adjacent monument, there is consanguinity. The effigies on our coins, the signs over shops, the figures that fill every ledger, the coats of arms outside the carriage panel, and the placards inside the omnibus, are, in common with dolls, blue-books, paper-hangings, lineally descended from the rude sculpture-paintings in which the Egyptians represented the triumphs and worship of their god-kings. Perhaps no example can be given which more vividly illustrates the multiplicity and heterogeneity of the products that in course of time may arise by successive differentiations from a common stock.

Before passing to other classes of facts, it should be observed that the evolution of the homogeneous into the heterogeneous is displayed not only in the separation of Painting and Sculpture from Architecture and from each other, and in the greater variety of subjects they embody but it is further shown in the structure of each work. A modern picture or statue is of far more heterogeneous nature than an ancient one. An Egyptian sculpture-fresco represents all its figures as on one plane — that is, at the game distance from the eye; and so is less heterogeneous than a painting that represents them as at various distances from the eye. It exhibits all objects as exposed to the game degree of light; and so is less heterogeneous than a painting which exhibits different objects and different parts of each object as in different degrees of light. It uses scarcely any but the primary colours, and these in their full intensity; and so is less heterogeneous than a painting which, introducing the primary colours but sparingly, employs an endless variety of intermediate tints, each of heterogeneous composition, and differing from the rest not only in quality but in intensity. Moreover, we see in these earliest works a great uniformity of conception. The same arrangement of figures is perpetually reproduced — the same actions, attitudes, faces, dresses. In Egypt the modes of representation were so fixed that it was sacrilege to introduce a novelty; and indeed it could have been only in consequence of a fixed mode of representation that a system, of hieroglyphics became possible. The Assyrian bas-reliefs display parallel characters. Deities, kings, attendants, winged figures and animals, are severally depicted in like positions, holding like implements, doing like things, and with like expression or non-expression of face. If a palm-grove is introduced, all the trees are of the same height, have the same number of leaves, and are equidistant. When water is imitated, each wave is a counterpart of the rest; and the fish, almost always of one kind, are evenly distributed over the surface. The beards of the kings, the gods, and the winged figures, are everywhere similar: as are the manes of the lions, and equally so those of the horses. Hair is represented throughout by one form of curl. The king's beard is quite architecturally built up of compound tiers of uniform curls, alternating with twisted tiers placed in a transverse direction, and arranged with perfect regularity; and the terminal tufts of the bulls' tails are represented in exactly the same manner. Without tracing out analogous facts in early Christian art, in which, though less striking, they are still visible, the advance in heterogeneity will be sufficiently manifest on remembering that in the pictures of our own day the composition is endlessly varied; the attitudes, faces, expressions, unlike; the subordinate objects different in size, form, position, texture; and more or less of contrast even in the smallest details. Or, if we compare an Egyptian statue, seated bolt upright on a block, with hands on knees, fingers outspread and parallel, eyes looking straight forward, and the two sides perfectly symmetrical in every particular, with a statue of the advanced Greek or the modern school, which is asymmetrical in respect of the position of the head, the body, the limbs, the arrangement of the hair, dress, appendages, and in its relations to neighbouring objects, we shall see the change from the homogeneous to the heterogeneous clearly manifested.

In the co-ordinate origin and gradual differentiation of Poetry, Music and Dancing, we have another series of illustrations. Rhythm in speech, rhythm in sound, and rhythm in motion, were in the beginning parts of the same thing, and have only in process of time become separate things. Among various existing barbarous tribes we find them still united. The dances of savages are accompanied by some kind of monotonous chant, the clapping of hands, the striking of rude instruments: there are measured movements, measured words, and measured tones; and the whole ceremony, usually having reference to war or sacrifice, is of governmental character. In the early records of the historic races we similarly find these three forms of metrical action united in religious festivals. In the Hebrew writings we read that the triumphal ode composed by Moses on the defeat of the Egyptians, was sung to an accompaniment of dancing and timbrels. The Israelites danced and sung "at the inauguration of the golden calf. And as it is generally agreed that this representation of the Deity was borrowed from the mysteries of Apis, it is probable that the dancing was copied from that of the Egyptians on those occasions." There was an annual dance in Shiloh on the sacred festival; and David danced before the ark. Again, in Greece the like relation is everywhere seen: the original type being there, as probably in other cases, a simultaneous chanting and mimetic representation of the life and adventures of the god. The Spartan dances were accompanied by hymns and songs; and in general the Greeks had "no festivals or religious assemblies but what were accompanied with songs and dances" — both of them being forms of worship used before altars. Among the Romans, too, there were sacred dances: the Salian and Lupercalian being named as of that kind. And even in Christian countries, as at Limoges, in comparatively recent times, the people have danced in the choir in honour of a saint. The incipient separation of these once united arts from each other and from religion, was early visible in Greece. Probably diverging from dances partly religious, partly warlike, as the Corybantian, came the war dances proper, of which there were various kinds; and from these resulted secular dances. Meanwhile Music and Poetry, though still united, came to have an existence separate from dancing. The aboriginal Greek poems, religious in subject, were not recited, but chanted; and though at first the chant of the poet was accompanied by the dance of the chorus, it ultimately grew into independence. Later still, when the poem had been differentiated into epic and lyric — when it became the custom to sing the lyric and recite the epic — poetry proper was born. As during the same period musical instruments were being multiplied, we may presume that music came to have an existence apart from words. And both of them were beginning to assume other forms besides the religious. Facts having like implications might be cited from the histories of later times and peoples: as the practices of our own early minstrels, who sang to the harp heroic narratives versified by themselves to music of their own composition; thus uniting the now separate offices of poet, composer, vocalist, and instrumentalist. But, without further illustration, the common origin and gradual differentiation of Dancing, Poetry, and Music will be sufficiently manifest.

The advance from the homogeneous to the heterogeneous is displayed not only in the separation of these arts from each other and from religion, but also in the multiplied differentiations which each of them afterwards undergoes. Not to dwell upon the numberless kinds of dancing that have, in course of time, come into use; and not to occupy space in detailing the progress of poetry, as seen in the development of the various forms of metre, of rhyme, and of general organization; let us confine our attention to music as a type of the group. As argued by Dr. Burney, and as implied by the customs of still extant barbarous races, the first musical instruments were, without doubt, percussive — sticks, calabashes, tom-toms — and were used simply to mark the time of the dance; and in this constant repetition of the same sound, we see music in its most homogeneous form.

The Egyptians had a lyre with three strings. The early lyre of the Greeks had four, constituting their tetra-chord. In course of some centuries lyres of seven and eight strings were employed. And, by the expiration of a thousand years, they had advanced to their "great system" of the double octave. Through all which changes there of course arose a greater heterogeneity of melody. Simultaneously there came into use the different modes — Dorian, Ionian, Phrygian, Æolian, and Lydian — answering to our keys; and of these there were ultimately fifteen. As yet, however, there was but little heterogeneity in the time of their music.

Instrumental music during this period being merely the accompaniment of vocal music, and vocal music being completely subordinated to words, the singer being also the poet, chanting his own compositions and making the lengths of his notes agree with the feet of his verses, — there unavoidably arose a tiresome uniformity of measure, which, as Dr. Burney says, "no resources of melody could disguise." Lacking the complex rhythm obtained by our equal bars and unequal notes the only rhythm was that produced by the quantity of the syllables and was of necessity comparatively monotonous. And further, it may be observed that the chant thus resulting, being like recitative, was much less clearly differentiated from ordinary speech than is our modern song.

Nevertheless, in virtue of the extended range of notes in use, the variety of modes, the occasional variations of time consequent on changes of metre, and the multiplication of instruments, music had, towards the close of Greek civilization, attained to considerable heterogeneity — not indeed as compared with our music, but as compared with that which preceded it. As yet, however, there existed nothing but melody: harmony was unknown. It was not until Christian church-music had reached some development, that music in parts was evolved; and then it came into existence through a very unobtrusive differentiation. Difficult as it may be to conceive *à priori* how the advance from melody to harmony could take place without a sudden leap, it is none the less true that it did so. The circumstance which prepared the way for it was the employment of two choirs singing alternately the same air. Afterwards it became the practice — very possibly first suggested by a mistake — for the second choir to commence before the first had ceased; thus producing fugue.

With the simple airs then in use, a partially harmonious fugue might not improbably thus result: and a very partially harmonious fugue satisfied the ears of that age, as we know from still preserved examples. The idea having once been given, the composing of airs productive of fugal harmony would naturally grow up; as in some way it *did* grow up out of this alternate choir-singing. And from the fugue to concerted music of two, three, four, and more parts, the transition was easy. Without pointing out in detail the increasing complexity that resulted from introducing notes of various lengths, from the multiplication of keys, from the use of accidentals, from varieties of time, and so forth, it needs but to contrast music as it is, with music as it was, to see how immense is the increase of heterogeneity. We see this if, looking at music in its *ensemble*, we enumerate its many different genera and species — if we consider the divisions into vocal, instrumental, and mixed; and their subdivisions into music for different voices and different instruments — if we observe the many forms of sacred music, from the simple hymn, the chant, the canon, motet, anthem, &c., up to the oratorio; and the still more numerous forms of secular music, from the ballad up to the serenata, from the instrumental solo up to the symphony.

Again, the same truth is seen on comparing any one sample of aboriginal music with a sample of modern music — even an ordinary song for the piano; which we find to be relatively highly heterogeneous, not only in respect of the varieties in the pitch and in the length of the notes, the number of different notes sounding at the same instant in company with the voice, and the variations of strength with which they are sounded and sung, but in respect of the changes of key, the changes of time, the changes of *timbre* of the voice, and the many other modifications of expression. While between the old monotonous dance-chant and a grand opera of our own day, with its endless orchestral complexities and vocal combinations, the contrast in heterogeneity is so extreme that it seems scarcely credible that the one should have been the ancestor of the other.

Were they needed, many further illustrations might be cited. Going back to the early time when the deeds of the god-king, chanted and mimetically represented in dances round his altar, were further narrated in picture-writings on the walls of temples and palaces, and so constituted a rude literature, we might trace the development of Literature through phases in which, as in the Hebrew Scriptures, it presents in one work theology, cosmogony, history, biography, civil law, ethics, poetry; through other phases in which, as in the Iliad, the religious, martial, historical, the epic, dramatic, and lyric elements are similarly commingled; down to its present heterogeneous development, in which its divisions and subdivisions are so numerous and varied as to defy complete classification. Or we might trace out the evolution of Science; beginning with the era in which it was not yet differentiated from Art, and was, in union with Art, the handmaid of Religion; passing through the era in which the sciences were so few and rudimentary, as to be simultaneously cultivated by the same philosophers; and ending with the era in which the genera and species are so numerous that few can enumerate them, and no one can adequately grasp even one genus. Or we might do the like with Architecture, with the Drama, with Dress.

But doubtless the reader is already weary of illustrations; and our promise has been amply fulfilled. We believe we have shown beyond question, that that which the German physiologists have found to be the law of organic development, is the law of all development. The advance from the simple to the complex, through a process of successive differentiations, is seen alike in the earliest changes of the Universe to which we can reason our way back; and in the earliest changes which we can inductively establish; it is seen in the geologic and climatic evolution of the Earth, and of every single organism on its surface; it is seen in the evolution of Humanity, whether contemplated in the civilized individual, or in the aggregation of races; it is seen in the evolution of Society in respect alike of its political, its religious, and its economical organization; and it is seen in the evolution of all those endless concrete and abstract products of human activity which constitute the environment of our daily life. From the remotest past which Science can fathom, up to the novelties of yesterday, that in which Progress essentially consists, is the transformation of the homogeneous into the heterogeneous.

And now, from this uniformity of procedure, may we not infer some fundamental necessity whence it results? May we not rationally seek for some all-pervading principle which determines this all-pervading process of things? Does not the universality of the *law* imply a universal *cause*?

That we can fathom such cause, noumenally considered, is not to be supposed. To do this would be to solve that ultimate mystery which must ever transcend human intelligence. But it still may be possible for us to reduce the law of all Progress, above established, from the condition of an empirical generalization, to the condition of a rational generalization. Just as it was possible to interpret Kepler's laws as necessary consequences of the law of gravitation; so it may be possible to interpret this law of Progress, in its multiform manifestations, as the necessary consequence of some similarly universal principle. As gravitation was assignable as the *cause* of each of the groups of phenomena which Kepler formulated; so may some equally simple attribute of things be assignable as the cause of each of the groups of phenomena formulated in the foregoing pages. We may be able to affiliate all these varied and complex evolutions of the homogeneous into the heterogeneous, upon certain simple facts of immediate experience, which, in virtue of endless repetition, we regard as necessary.

The probability of a common cause, and the possibility of formulating it, being granted, it will be well, before going further, to consider what must be the general characteristics of such cause, and in what direction we ought to look for it. We can with certainty predict that it has a high degree of generality; seeing that it is common to such infinitely varied phenomena: just in proportion to the universality of its application must be the abstractness of its character. We need not expect to sea in it an obvious solution of this or that form of Progress; because it equally refers to forms of Progress bearing little apparent resemblance to them: its association with multiform orders of facts, involves its dissociation from any particular order of facts. Being that which determines Progress of every kind — astronomic, geologic, organic, ethnologic, social, economic, artistic, &c. — it must be concerned with some fundamental attribute possessed in common by these; and must be expressible in terms of this fundamental attribute. The only obvious respect in which all kinds of Progress are alike, is, that they are modes of change; and hence, in some characteristic of changes in general, the desired solution will probably be found. We may suspect *à priori* that in some law of change lies the explanation of this universal transformation of the homogeneous into the heterogeneous.

Thus much premised, we pass at once to the statement of the law, which is this: — *Every active force produces more than one change — every cause produces more than one effect.*

Before this law can be duly comprehended, a few examples mast be looked at. When one body is struck against another, that which we usually regard as the effect, is a change of position or motion in one or both bodies. But a moment's thought shows us that this is a careless and very incomplete view of the matter. Besides the visible mechanical result, sound is produced; or, to speak accurately, a vibration in one or both bodies, and in the surrounding air: and under some circumstances we call this the effect. Moreover, the air has not only been made to vibrate, but has had sundry currents caused in it by the transit of the bodies. Further, there is a disarrangement of the particles of the two bodies in the neighbourhood of their point of collision; amounting in some cases to a visible condensation. Yet more, this condensation is accompanied by the disengagement of heat. In some cases a spark — that is, light — results, from the incandescence of a portion struck off; and sometimes this incandescence is associated with chemical combination.

Thus, by the original mechanical force expended in the collision, at least five, and often more, different kinds of changes have been produced. Take, again, the lighting of a candle. Primarily this is a chemical change consequent on a rise of temperature. The process of combination having once been set going by extraneous heat, there is a continued formation of carbonic acid, water, &c. — in itself a result more complex than the extraneous heat that first caused it. But accompanying this process of combination there is a production of heat; there is a production of light; there is an ascending column of hot gases generated; there are currents established in the surrounding air. Moreover the decomposition of one force into many forces does not end here: each of the several changes produced becomes the parent of further changes. The carbonic acid given off will by and by combine with some base; or under the influence of sunshine give up its carbon to the leaf of a plant. The water will modify the hygrometric state of the air around; or, if the current of hot gases containing it come against a cold body, will be condensed: altering the temperature, and perhaps the chemical state, of the surface it covers. The heat given out melts the subjacent tallow, and expands whatever it warms. The light, falling on various substances, calls forth from them reactions by which it is modified; and so divers colours are produced. Similarly even with these secondary actions, which may be traced out into ever-multiplying ramifications, until they become too minute to be appreciated. And thus it is with all changes whatever. No case can be named in which an active force does not evolve forces of several kinds, and each of these, other groups of forces. Universally the effect is more complex than the cause.

Doubtless the reader already foresees the course of our argument. This multiplication of results, which is displayed in every event of to-day, has been going on from the beginning; and is true of the grandest phenomena of the universe as of the most insignificant. From the law that every active force produces more than one change, it is an inevitable corollary that through all time there has been an ever-growing complication of things. Starting with the ultimate fact that every cause produces more than one effect, we may readily see that throughout creation there must have gone on, and must still go on, a never-ceasing transformation of the homogeneous into the heterogeneous. But let us trace out this truth in detail.[2]

Without committing ourselves to it as more than a speculation, though a highly probable one, let us again commence with the evolution of the solar system out of a nebulous medium.[3] From the mutual attraction of the atoms of a diffused mass whose form is unsymmetrical, there results not only condensation but rotation: gravitation simultaneously generates both the centripetal and the centrifugal forces. While the condensation and the rate of rotation are progressively increasing, the approach of the atoms necessarily generates a progressively increasing temperature. As this temperature rises, light begins to be evolved; and ultimately there results a revolving sphere of fluid matter radiating intense heat and light — a sun.

There are good reasons for believing that, in consequence of the high tangential velocity, and consequent centrifugal force, acquired by the outer parts of the condensing nebulous mass, there must be a periodical detachment of rotating rings; and that, from the breaking up of these nebulous rings, there must arise masses which in the course of their condensation repeat the actions of the parent mass, and so produce planets and their satellites — an inference strongly supported by the still extant rings of Saturn.

Should it hereafter be satisfactorily shown that planets and satellites were thus generated, a striking illustration will be afforded of the highly heterogeneous effects produced by the primary homogeneous cause; but it will serve our present purpose to point to the fact that from the mutual attraction of the particles of an irregular nebulous mass there result condensation, rotation, heat, and light.

It follows as a corollary from the Nebular Hypothesis, that

---

[2] A correlative truth which ought also to be taken into account (that the state of homogeneity is one of unstable equilibrium), but which it would greatly encumber the argument to exemplify in connection with the above, will be found developed in the essay on *Transcendental Physiology*.

[3] The idea that the Nebular Hypothesis has been disproved because what were thought to be existing nebulas have been resolved into clusters of stars is almost beneath notice. *A priori* it was highly improbable, if not impossible, that nebulous masses should still remain uncondensed, while others have been condensed millions of years ago.

the Earth must at first have been incandescent; and whether the Nebular Hypothesis be true or not, this original incandescence of the Earth is now inductively established — or, if not established, at least rendered so highly probable that it is a generally admitted geological doctrine. Let us look first at the astronomical attributes of this once molten globe. From its rotation there result the oblateness of its form, the alternations of day and night, and (under the influence of the moon) the tides, aqueous and atmospheric. From the inclination of its axis, there result the precession of the equinoxes and the many differences of the seasons, both simultaneous and successive, that pervade its surface. Thus the multiplication of effects is obvious Several of the differentiations due to the gradual cooling of the Earth have been already noticed — as the formation of a crust, the solidification of sublimed elements, the precipitation of water, &c., — and we here again refer to them merely to point out that they are simultaneous effects of the one cause, diminishing heat.

Let us now, however, observe the multiplied changes afterwards arising from the continuance of this one cause. The cooling of the Earth involves its contraction. Hence the solid crust first formed is presently too large for the shrinking nucleus; and as it cannot support itself, inevitably follows the nucleus. But a spheroidal envelope cannot sink down into contact with a smaller internal spheroid, without disruption; it must run into wrinkles as the rind of an apple does when the bulk of its interior decreases from evaporation. As the cooling progresses and the envelope thickens, the ridges consequent on these contractions must become greater, rising ultimately into hills and mountains; and the later systems of mountains thus produced must not only be higher, as we find them to be, but they must be longer, as we also find them to be. Thus, leaving out of view other modifying forces, we see what immense heterogeneity of surface has arisen from the one cause, loss of heat — a heterogeneity which the telescope shows us to be paralleled on the face of the moon, where aqueous and atmospheric agencies have been absent.

But we have yet to notice another kind of heterogeneity of surface similarly and simultaneously caused. While the Earth's crust was still thin, the ridges produced by its contraction must not only have been small, but the spaces between these ridges must have rested with great evenness upon the subjacent liquid spheroid; and the water in those arctic and antarctic regions in which it first condensed, must have been evenly distributed. But as fast as the crust grew thicker and gained corresponding strength, the lines of fracture from time to time caused in it, must have occurred at greater distances apart; the intermediate surfaces must have followed the contracting nucleus with less uniformity; and there must have resulted larger areas of land and water. If any one, after wrapping up an orange in wet tissue paper, and observing not only how small are the wrinkles, but how evenly the intervening spaces lie upon the surface of the orange, will then wrap it up in thick cartridge-paper, and note both the greater height of the ridges and the much larger spaces throughout which the paper does not touch the orange, he will realize the fact, that as the Earth's solid envelope grew thicker, the areas of elevation and depression must have become greater. In place of islands more or less homogeneously scattered over an all-embracing sea, there must have gradually arisen heterogeneous arrangements of continent and ocean, such as we now know.

Once more, this double change in the extent and in the elevation of the lands, involved yet another species of heterogeneity, that of coast-line. A tolerably even surface raised out of the ocean, must have a simple, regular sea-margin; but a surface varied by table-lands and intersected by mountain-chains must, when raised out of the ocean, have an outline extremely irregular both in its leading features and in its details. Thus endless is the accumulation of geological and geographical results slowly brought about by this one cause — the contraction of the Earth.

When we pass from the agency which geologists term igneous, to aqueous and atmospheric agencies, we see the like ever-growing complications of effects. The denuding actions of air and water have, from the beginning, been modifying every exposed surface; everywhere causing many different changes. Oxidation, heat, wind, frost, rain, glaciers, rivers, tides, waves, have been unceasingly producing disintegration; varying in kind and amount according to local circumstances. Acting upon a tract of granite, they here work scarcely an appreciable effect; there cause exfoliations of the surface, and a resulting heap of *débris* and boulders; and elsewhere, after decomposing the feldspar into a white clay, carry away this and the accompanying quartz and mica, and deposite them in separate beds, fluviatile and marine.

When the exposed land consists of several unlike formations, sedimentary and igneous, the denudation produces changes proportionably more heterogeneous. The formations being disintegrable in different degrees, there follows an increased irregularity of surface. The areas drained by different rivers being differently constituted, these rivers carry down to the sea different combinations of ingredients; and so sundry new strata of distinct composition are formed.

And here indeed we may see very simply illustrated, the truth, which we shall presently have to trace out in more involved cases, that in proportion to the heterogeneity of the object or objects on which any force expends itself, is the heterogeneity of the results. A continent of complex structure, exposing many strata irregularly distributed, raised to various levels, tilted up at all angles, must, under the same denuding agencies, give origin to immensely multiplied results; each district must be differently modified; each river must carry down a different kind of detritus; each deposit must be differently distributed by the entangled currents, tidal and other, which wash the contorted shores; and this multiplication of results must manifestly be greatest where the complexity of the surface is greatest.

It is out of the question here to trace in detail the genesis of those endless complications described by Geology and Physical Geography: else we might show how the general truth, that every active force produces more than one change, is exemplified in the highly involved flow of the tides, in the ocean currents, in the winds, in the distribution of rain, in the distribution of heat, and so forth. But not to dwell upon these, let us, for the fuller elucidation of this truth in relation to the inorganic world, consider what would be the consequences of some extensive cosmical revolution — say the subsidence of Central America.

The immediate results of the disturbance would themselves be sufficiently complex. Besides the numberless dislocations of strata, the ejections of igneous matter, the propagation of earthquake vibrations thousands of miles around, the loud explosions, and the escape of gases; there would be the rush of the Atlantic and Pacific Oceans to supply the vacant space, the subsequent recoil of enormous waves, which would traverse both these oceans and produce myriads of changes along their shores, the corresponding atmospheric waves complicated by the currents surrounding each volcanic vent, and the electrical discharges with which much disturbances are accompanied. But these temporary effects would be insignificant compared with the permanent ones. The complex currents of the Atlantic and Pacific would be altered in direction and amount. The distribution of heat achieved by these ocean currents would be different from what it is. The arrangement of the isothermal lines, not even on the neighbouring continents, but even throughout Europe, would be changed. The tides would flow differently from what they do now. There would be more or less modification of the winds in their periods, strengths, directions, qualities. Rain would fall scarcely anywhere at the same times and in the same quantities as at present. In short, the meteorological conditions thousands of miles off, on all sides, would be more or less revolutionized.

Thus, without taking into account the infinitude of modifications which these changes of climate would produce upon the flora and fauna, both of land and sea, the reader will see the immense heterogeneity of the results wrought out by one force, when that force expends itself upon a previously complicated area; and he will readily draw the corollary that from the beginning the complication has advanced at an increasing rate.

Before going on to show how organic progress also depends upon the universal law that every force produces more than one change, we have to notice the manifestation of this law in yet another species of inorganic progress — namely, chemical. The same general causes that have wrought out the heterogeneity of the Earth, physically considered, have simultaneously wrought out its chemical heterogeneity. Without dwelling upon the general fact that the forces which have been increasing the variety and complexity of geological formations, have, at the same time been bringing into contact elements not previously exposed to each other under conditions favourable to union, and so have been adding to the number of chemical compounds, let us pass to the more important complications that have resulted from the cooling of the Earth.

There is every reason to believe that at an extreme heat the elements cannot combine. Even under such heat as can be artificially produced, some very strong affinities yield, as for instance, that of oxygen for hydrogen; and the great majority of chemical compounds are decomposed at much lower temperatures. But without insisting upon the highly probable inference, that when the Earth was in its first state of incandescence there were no chemical combinations at all, it will suffice our purpose to point to the unquestionable fact that the compounds that can exist at the highest temperatures, and which must, therefore, have been the first that were formed as the Earth cooled, am those of the simplest constitutions. The protoxides — including under that head the alkalies, earths, &c. — are, as a class, the most stable compounds we know: most of them resisting decomposition by any heat we can generate. These, consisting severally of one atom of each component element, are combinations of the simplest order — are but one degree less homogeneous than the elements themselves. More heterogeneous than these, less stable, and therefore later in the Earth's history, are the deutoxides, tritoxides, peroxides, &c.; in which two, three, four, or more atoms of oxygen are united with one atom of metal or other element. Higher than these in heterogeneity are the hydrates; in which an oxide of hydrogen, united with an oxide of some other element, forms a substance whose atoms severally contain at least four ultimate atoms of three different kinds. Yet more heterogeneous and less stable still are the salts; which present us with compound atoms each made up of five, six, seven, eight, ten, twelve, or more atoms, of three, if not more, kinds. Then there are the hydrated salts, of a yet greater heterogeneity, which undergo partial decomposition at much lower temperatures. After them come the further-complicated supersalts and double salts, having a stability again decreased: and so throughout. Without entering into qualifications for which we lack space, we believe no chemist will deny it to be a general law of these inorganic combinations that, *other things equal*, the stability decreases as the complexity increases.

And then when we pass to the compounds of organic chemistry, we find this general law still further exemplified: we find much greater complexity and much less stability. An atom of albumen, for instance, consists of 482 ultimate atoms of five different kinds. Fibrine, still more intricate in constitution, contains in each atom, 298 atoms of carbon, 40 of nitrogen, 2 of sulphur, 228 of hydrogen, and 92 of oxygen — in all, 600 atoms; or, more strictly speaking — equivalents. And these two substances are so unstable as to decompose at quite ordinary temperatures; as that to which the outside of a joint of roast meat is exposed. Thus it is manifest that the present chemical heterogeneity of the Earth's surface has arisen by degrees, as the decrease of heat has permitted; and that it has shown itself in three forms — first, in the multiplication of chemical compounds; second, in the greater number of different elements contained in the more modern of these compounds: and third, in the higher and more varied multiples in which these more numerous elements combine.

To say that this advance in chemical heterogeneity is due to the one cause, diminution of the Earth's temperature, would be to say too much; for it is clear that aqueous and atmospheric agencies have been concerned; and, further, that the affinities of the elements themselves are implied. The cause has all along been a composite one: the cooling of the Earth having been simply the most general of the concurrent causes, or assemblage of conditions. And here, indeed, it may be remarked that in the several classes of facts already dealt with (excepting, perhaps, the first), and still more in those with which we shall presently deal, the causes are more or less compound; as indeed are nearly all causes with which we are acquainted. Scarcely any change can with logical accuracy be wholly ascribed to one agency, to the neglect of the permanent or temporary conditions under which only this agency produces the change. But as it does not materially affect our argument, we prefer, for simplicity's sake, to use throughout the popular mode of expression.

Perhaps it will be further objected, that to assign loss of heat as the cause of any changes, is to attribute these changes not to a force, but to the absence of a force. And this is true. Strictly speaking, the changes should be attributed to those forces which come into action when the antagonist force is withdrawn. But though there is an inaccuracy in saying that the freezing of water is due to the loss of its heat, no practical error arises from it; nor will a parallel laxity of expression vitiate our statements respecting the multiplication of effects. Indeed, the objection serves but to draw attention to the fact, that not only does the exertion of a force produce more than one change, but the withdrawal of a force produces more than one change. And this suggests that perhaps the most correct statement of our general principle would be its most abstract statement — every change is followed by more than one other change.

Returning to the thread of our exposition, we have next to trace out, in organic progress, this same all-pervading principle. And here, where the evolution of the homogeneous into the heterogeneous was first observed, the production of many changes by one cause is least easy to demonstrate. The development of a seed into a plant, or an ovum into an animal, is so gradual, while the forces which determine it are so involved, and at the same time so unobtrusive, that it is difficult to detect the multiplication of effects which is elsewhere so obvious. Nevertheless, guided by indirect evidence, we may pretty safely reach the conclusion that here too the law holds.

Observe, first, how numerous are the effects which any marked change works upon an adult organism — a human being, for instance. An alarming sound or sight, besides the impressions on the organs of sense and the nerves, may produce a start, a scream, a distortion of the face, a trembling consequent upon a general muscular relaxation, a burst of perspiration, an excited action of the heart, a rush of blood to the brain, followed possibly by arrest of the heart's action and by syncope: and if the system be feeble, an indisposition with its long train of complicated symptoms may set in. Similarly in cases of disease. A minute portion of the small-pox virus introduced into the system, will, in a severe case, cause, during the first stage, rigors, heat of skin, accelerated pulse, furred tongue, loss of appetite, thirst, epigastric uneasiness, vomiting, headache, pains in the back and limbs, muscular weakness, convulsions, delirium, &c.; in the second stage, cutaneous eruption, itching, tingling, sore throat, swelled fauces, salivation, cough, hoarseness, dyspnœa, &c.; and in the third stage, œdematous inflammations, pneumonia, pleurisy, diarrhœa, inflammation of the brain, ophthalmia, erysipelas, &c.: each of which enumerated symptoms is itself more or less complex. Medicines, special foods, better air, might in like manner be instanced as producing multiplied results.

Now it needs only to consider that the many changes thus wrought by one force upon an adult organism, will be in part paralleled in an embryo organism, to understand how here also, the evolution of the homogeneous into the heterogeneous may be due to the production of many effects by one cause. The external heat and other agencies which determine the first complications of the germ, may, by acting upon these, superinduce further complications; upon these still higher and more numerous ones; and so on continually: each organ as it is developed serving, by its actions and reactions upon the rest, to initiate new complexities. The first pulsations of the fœtal heart must simultaneously aid the unfolding of every part. The growth of each tissue, by taking from the blood special proportions of elements, must modify the constitution of the blood; and so must modify the nutrition of all the other tissues. The heart's action, implying as it does a certain waste, necessitates an addition to the blood of effete matters, which must influence the rest of the system, and perhaps, as some think, cause the formation of excretory organs. The nervous connections established among the viscera must further multiply their mutual influences: and so continually.

Still stronger becomes the probability of this view when we call to mind the fact, that the same germ may be evolved into different forms according to circumstances. Thus, during its earlier stages, every embryo is sexless — becomes either male or female as the balance of forces acting upon it determines. Again, it is a well-established fact that the larva of a working-bee will develop into a queen-bee, if, before it is too late, its food be changed to that on which the larvae of queen-bees are fed. Even more remarkable is the case of certain entozoa. The ovum of a tape-worm, getting into its natural habitat, the intestine, unfolds into the well-known form of its parent; but if carried, as it frequently is, into other parts of the system, it becomes a sac-like creature, called by naturalists the *Echinococcus* — a creature so extremely different from the tape-worm in aspect and structure, that only after careful investigations has it been proved to have the same origin. All which instances imply that each advance in embryonic complication results from the action of incident forces upon the complication previously existing.

Indeed, we may find *à priori* reason to think that the evolution proceeds after this manner. For since it is now known that no germ, animal or vegetable, contains the slightest rudiment, trace, or indication of the future organism — now that the microscope has shown us that the first process set up in every fertilized germ, is a process of repeated spontaneous fissions ending in the production of a mass of cells, not one of which exhibits any special character: there seems no alternative but to suppose that the partial organization at any moment subsisting in a growing embryo, is transformed by the agencies acting upon it into the succeeding phase of organization, and this into the next, until, through ever-increasing complexities, the ultimate form is reached. Thus, though the subtilty of the forces and the slowness of the results, prevent us from *directly* showing that the stages of increasing heterogeneity through which every embryo passes, severally arise from the production of many changes by one force, yet, *indirectly*, we have strong evidence that they do so.

We have marked how multitudinous are the effects which one cause may generate in an adult organism; that a like multiplication of effects must happen in the unfolding organism, we have observed in sundry illustrative cases; further, it has been pointed out that the ability which like germs have to originate unlike forms, implies that the successive transformations result from the new changes superinduced on previous changes; and we have seen that structureless as every germ originally is, the development of an organism out of it is otherwise incomprehensible. Not indeed that we can thus really explain the production of any plant or animal. We are still in the dark respecting those mysterious properties in virtue of which the germ, when subject to fit influences, undergoes the special changes that begin the series of transformations. All we aim to show, is, that given a germ possessing these mysterious properties, the evolution of an organism from it, probably depends upon that multiplication of effects which we have seen to be the cause of progress in general, so far as we have yet traced it.

When, leaving the development of single plants and animals, we pass to that of the Earth's flora and fauna, the course of our argument again becomes clear and simple. Though, as was admitted in the first part of this article, the fragmentary facts Palæontology has accumulated, do not clearly warrant us in saying that, in the lapse of geologic time, there have been evolved more heterogeneous organisms, and more heterogeneous assemblages of organisms, yet we shall now see that there *must* ever have been a tendency towards these results. We shall find that the production of many effects by one cause, which, as already shown, has been all along increasing the physical heterogeneity of the Earth, has further involved an increasing heterogeneity in its flora and fauna, individually and collectively. An illustration will make this clear.

Suppose that by a series of upheavals, occurring, as they are now known to do, at long intervals, the East Indian Archipelago were to be, step by step, raised into a continent, and a chain of mountains formed along the axis of elevation. By the first of these upheavals, the plants and animals inhabiting Borneo, Sumatra, New Guinea, and the rest, would be subjected to slightly modified sets of conditions. The climate in general would be altered in temperature, in humidity, and in its periodical variations; while the local differences would be multiplied. These modifications would affect, perhaps inappreciably, the entire flora and fauna of the region. The change of level would produce additional modifications: varying in different species, and also in different members of the same species, according to their distance from the axis of elevation. Plants, growing only on the sea-shore in special localities, might become extinct. Others, living only in swamps of a certain humidity, would, if they survived at all, probably undergo visible changes of appearance. While still greater alterations would occur in the plants gradually spreading over the lands newly raised above the sea. The animals and insects living on these modified plants, would themselves be in some degree modified by change of food, as well as by change of climate; and the modification would be more marked where, from the dwindling or disappearance of one kind of plant, an allied kind was eaten. In the lapse of the many generations arising before the next upheaval, the sensible or insensible alterations thus produced in each species would become organized — there would be a more or less complete adaptation to the new conditions. The next upheaval would superinduce further organic changes, implying wider divergences from the primary forms; and so repeatedly.

But now let it be observed that the revolution thus resulting would not be a substitution of a thousand more or less modified species for the thousand original species; but in place of the thousand original species there would arise several thousand species, or varieties, or changed forms. Each species being distributed over an area of some extent, and tending continually to colonize the new area exposed, its different members would be subject to different sets of changes. Plants and animals spreading towards the equator would not be affected in the same way with others spreading from it. Those spreading towards the new shores would undergo changes unlike the changes undergone by those spreading into the mountains. Thus, each original race of organisms, would become the root from which diverged several races differing more or less from it and from each other; and while some of these might subsequently disappear, probably more than one would survive in the next geologic period: the very dispersion itself increasing the chances of survival. Not only would there be certain modifications thus caused by change of physical conditions and food, but also in some cases other modifications caused by change of habit. The fauna of each island, peopling, step by step, the newly-raised tracts, would eventually come in contact with the faunas of other islands; and some members of these other faunas would be unlike any creatures before seen. Herbivores meeting with new beasts of prey, would, in some cases, be led into modes of defence or escape differing from those previously used; and simultaneously the beasts of prey would modify their modes of pursuit and attack. We know that when circumstances demand it, such changes of habit *do* take place in animals; and we know that if the new habits become the dominant ones, they must eventually in some degree alter the organization.

Observe, now, however, a further consequence. There must arise not simply a tendency towards the differentiation of each race of organisms into several races; but also a tendency to the occasional production of a somewhat higher organism. Taken in the mass these divergent varieties which have been caused by fresh physical conditions and habits of life, will exhibit changes quite indefinite in kind and degree; and changes that do not necessarily constitute an advance. Probably in most cases the modified type will be neither more nor less heterogeneous than the original one. In some cases the habits of life adopted being simpler than before, a less heterogeneous structure will result: there will be a retrogradation. But it *must* now and then occur, that some division of a species, falling into circumstances which give it rather more complex experiences, and demand actions somewhat more involved, will have certain of its organs further differentiated in proportionately small degrees, — will become slightly more heterogeneous.

Thus, in the natural course of things, there will from time to time arise an increased heterogeneity both of the Earth's flora and fauna, and of individual races included in them. Omitting detailed explanations, and allowing for the qualifications which cannot here be specified, we think it is clear that geological mutations have all along tended to complicate the forms of life, whether regarded separately or collectively. The same causes which have led to the evolution of the Earth's crust from the simple into the complex, have simultaneously led to a parallel evolution of the Life upon its surface. In this case, as in previous ones, we see that the transformation of the homogeneous into the heterogeneous is consequent upon the universal principle, that every active force produces more than one change.

The deduction here drawn from the established truths of geology and the general laws of life, gains immensely in weight on finding it to be in harmony with an induction drawn from direct experience. Just that divergence of many races from one race, which we inferred must have been continually occurring during geologic time, we know to have occurred during the pre-historic and historic periods, in man and domestic animals. And just that multiplication of effects which we concluded must have produced the first, we see has produced the last. Single causes, as famine, pressure of population, war, have periodically led to further dispersions of mankind and of dependent creatures: each such dispersion initiating new modifications, new varieties of type. Whether all the human races be or be not derived from one stock, philology makes it clear that whole groups of races now easily distinguishable from each other, were originally one race, — that the diffusion of one race into different climates and conditions of existence, has produced many modified forms of it.

Similarly with domestic animals. Though in some cases — as that of dogs — community of origin will perhaps be disputed, yet in other cases — as that of the sheep or the cattle of our own country — it will not be questioned that local differences of climate, food, and treatment, have transformed one original breed into numerous breeds now become so far distinct as to produce unstable hybrids. Moreover, through the complications of effects flowing from single causes, we here find, what we before inferred, not only an increase of general heterogeneity, but also of special heterogeneity. While of the divergent divisions and subdivisions of the human race, many have undergone changes not constituting an advance; while in some the type may have degraded; in others it has become decidedly more heterogeneous. The civilized European departs more widely from the vertebrate archetype than does the savage. Thus, both the law and the cause of progress, which, from lack of evidence, can be but hypothetically substantiated in respect of the earlier forms of life on our globe, can be actually substantiated in respect of the latest forms.

If the advance of Man towards greater heterogeneity is traceable to the production of many effects by one cause, still more clearly may the advance of Society towards greater heterogeneity be so explained. Consider the growth of an industrial organization. When, as must occasionally happen, some individual of a tribe displays unusual aptitude for making an article of general use — a weapon, for instance — which was before made by each man for himself, there arises a tendency towards the differentiation of that individual into a maker of such weapon. His companions — warriors and hunters all of them, — severally feel the importance of having the best weapons that can be made; and are therefore certain to offer strong inducements to this skilled individual to make weapons for them. He, on the other hand, having not only an unusual faculty, but an unusual liking, for making such weapons (the talent and the desire for any occupation being commonly associated), is predisposed to fulfil these commissions on the offer of an adequate reward: especially as his love of distinction is also gratified. This first specialization of function, once commenced, tends ever to become more decided. On the side of the weapon-maker continued practice gives increased skill — increased superiority to his products: on the side of his clients, cessation of practice entails decreased skill. Thus the influences that determine this division of labour grow stronger in both ways; and the incipient heterogeneity is, on the average of cases, likely to become permanent for that generation, if no longer.

Observe now, however, that this process not only differentiates the social mass into two parts, the one monopolizing, or almost monopolizing, the performance of a certain function, and the other having lost the habit, and in some measure the power, of performing that function; but it tends to imitate other differentiations. The advance we have described implies the introduction of barter, — the maker of weapons has, on each occasion, to be paid in such other articles as he agrees to take in exchange.

But he will not habitually take in exchange one kind of article, but many kinds. He does not want mats only, or skins, or fishing gear, but he wants all these; and on each occasion will bargain for the particular things he most needs. What follows? If among the members of the tribe there exist any slight differences of skill in the manufacture of these various things, as there are almost sure to do, the weapon-maker will take from each one the thing which that one excels in making: he will exchange for mats with him whose mats are superior, and will bargain for the fishing gear of whoever has the best. But he who has bartered away his mats or his fishing gear, must make other mats or fishing gear for himself; and in so doing must, in some degree, farther develop his aptitude. Thus it results that the small specialities of faculty possessed by various members of the tribe, will tend to grow more decided. If such transactions are from time repeated, these specializations may become appreciable. And whether or not there ensue distinct differentiations of other individuals into makers of particular articles, it is clear that incipient differentiations take place throughout the tribe: the one original cause produces not only the first dual effect, but a number of secondary dual effects, like in kind, but minor in degree. This process, of which traces may be seen among groups of schoolboys, cannot well produce any lasting effects is an unsettled tribe; but where there grows up a fixed and. multiplying community, these differentiations become permanent, and increase with each generation. A larger population, involving a greater demand for every commodity, intensifies the functional activity of each specialized person or class; and this renders the specialization more definite where it already exists, and establishes it where it is nascent. By increasing the pressure on the means of subsistence, a larger population again augments these results; seeing that each person is forced more and more to confine himself to that which he can do best, and by which he can gain most. This industrial progress, by aiding future production, opens the way for a further growth of population, which reacts as before: in all which the multiplication of effects is manifest. Presently, under these same stimuli, new occupations arise. Competing workers, ever aiming to produce improved articles, occasionally discover better processes or raw materials. In weapons and cutting tools, the substitution of bronze for stone entails upon him who first makes it a great increase of demand — so great an increase that he presently finds, all his time occupied in making the bronze for the articles he sells, and is obliged to depute the fashioning of these to others: and, eventually, the making of bronze, thus gradually differentiated from a pre-existing occupation, becomes an occupation by itself.

But now mark the ramified changes which follow this change. Bronze soon replaces stone, not only in the articles it was first used for, but in many others — in arms, tools, and utensils of various kinds; and so affects the manufacture of these things. Further, it affects the processes which these utensils subserve, and the resulting products — modifies buildings, carvings, dress, personal decorations. Yet again, it sets going sundry manufactures which were before impossible, from lack of a material fit for the requisite tools. And all these changes react on the people — increase their manipulative skill, their intelligence, their comfort, — refine their habits and tastes. Thus the evolution of a homogeneous society into a heterogeneous one, is clearly consequent on the general principle, that many effects are produced by one cause.

Our limits will not allow us to follow out this process in its higher complications: else might we show how the localization of special industries in special parts of a kingdom, as well as the minute subdivision of labour in the making of each commodity, are similarly determined. Or, turning to a somewhat different order of illustrations, we might dwell on the multitudinous changes — material, intellectual, moral, — caused by printing; or the further extensive series of changes wrought by gunpowder. But leaving the intermediate phases of social development, let us take a few illustrations from its most recent and its passing phases. To trace the effects of steam-power, in its manifold applications to mining, navigation, and manufactures of all kinds, would carry us into unmanageable detail. Let us confine ourselves to the latest embodiment of steam-power — the locomotive engine.

This, as the proximate cause of our railway system, has changed the face of the country, the course of trade, and the habits of the people. Consider, first, the complicated sets of changes that precede the making of every railway — the provisional arrangements, the meetings, the registration, the trial section, the parliamentary survey, the lithographed plans, the books of reference, the local deposits and notices, the application to Parliament, the passing Standing-Orders Committee, the first, second, and third readings: each of which brief heads indicates a multiplicity of transactions, and the development of sundry occupations — as those of engineers, surveyors, lithographers, parliamentary agents, share-brokers; and the creation of sundry others — as those of traffic-takers, reference-takers. Consider, next, the yet more marked changes implied in railway construction — the cuttings, embankings, tunnellings, diversions of roads; the building of bridges and stations; the laying down of ballast, sleepers, and rails; the making of engines, tenders, carriages and waggons: which processes, acting upon numerous trades, increase the importation of timber, the quarrying of stone, the manufacture of iron, the mining of coal, the burning of bricks: institute a variety of special manufactures weekly advertised in the *Railway Times;* and, finally, open the way to sundry new occupations, as those of drivers, stokers, cleaners, plate-layers, &c., &c. And then consider the changes, more numerous and involved still, which railways in action produce on the community at large. The organization of every business is more or less modified: ease of communication makes it better to do directly what was before done by proxy; agencies are established where previously they would not have paid; goods are obtained from remote wholesale houses instead of near retail ones; and commodities are used which distance once rendered inaccessible. Again, the rapidity and small cost of carriage tend to specialize more than ever the industries of different districts — to confine each manufacture to the parts in which, from local advantages, it can be best carried on. Further, the diminished cost of carriage, facilitating distribution, equalizes prices, and also, on the average, lowers prices: thus bringing divers articles within the means of those before unable to buy them, and so increasing their comforts and improving their habits. At the same time the practice of travelling is immensely extended. Classes who never before thought of it, take annual trips to the sea; visit their distant relations; make tours; and so we are benefited in body, feelings, and intellect. Moreover, the more prompt transmission of letters and of news produces further changes — makes the pulse of the nation faster. Yet more, there arises a wide dissemination of cheap literature through railway book-stalls, and of advertisements in railway carriages: both of them aiding ulterior progress.

And all the innumerable changes here briefly indicated are consequent on the invention of the locomotive engine. The social organism has been rendered more heterogeneous in virtue of the many new occupations introduced, and the many old ones further specialized; prices in every place have been altered; each trader has, more or less, modified his way of doing business; and almost every person has been affected in his actions, thoughts, emotions.

Illustrations to the same effect might be indefinitely accumulated. That every influence brought to bear upon society works multiplied effects; and that increase of heterogeneity is due to this multiplication of effects; may be seen in the history of every trade, every custom, every belief. But it is needless to give additional evidence of this. The only further fact demanding notice, is, that we here see still more clearly than ever, the truth before pointed out, that in proportion as the area on which any force expends itself becomes heterogeneous, the results are in a yet higher degree multiplied in number and kind. While among the primitive tribes to whom it was first known, caoutchouc caused but a few changes, among ourselves the changes have been so many and varied that the history of them occupies a volume.[4] Upon the small, homogeneous community inhabiting one of the Hebrides, the electric telegraph would produce, were it used, scarcely any results; but is England the results it produces are multitudinous. The comparatively simple organization under which our ancestors lived five centuries ago, could have undergone but few modifications from an event like the recent one at Canton; but now the legislative decision respecting it sets up many hundreds of complex modifications, each of which will be

---

4 "Personal Narrative of the Origin of the Caoutchouc, or India-Rubber Manufacture in England." By Thomas Hancock.

the parent of numerous future ones.

Space permitting, we could willingly have pursued the argument in relation to all the subtler results of civilization. As before, we showed that the law of Progress to which the organic and inorganic worlds conform, is also conformed to by Language, Sculpture, Music, &c.; so might we here show that the cause which we have hitherto found to determine Progress holds in these cases also. We might demonstrate in detail how, in Science, an advance of one division presently advances other divisions — how Astronomy has been immensely forwarded by discoveries in Optics, while other optical discoveries have initiated Microscopic Anatomy, and greatly aided the growth of Physiology — how Chemistry has indirectly increased our knowledge of Electricity, Magnetism, Biology, Geology — how Electricity has reacted on Chemistry and Magnetism, developed our views of Light and Heat, and disclosed sundry laws of nervous action.

In Literature the same truth might be exhibited in the manifold effects of the primitive mystery-play, not only as originating the modern drama, but as affecting through it ether kinds of poetry and fiction; or in the still multiplying forms of periodical literature that have descended from the first newspaper, and which have severally acted and reacted on other forms of literature and on each other. The influence which a new school of Painting — as that of the pre-Raffaelites — exercises upon other schools; the hints which all kinds of pictorial art are deriving from Photography; the complex results of new critical doctrines, as those of Mr. Ruskin, might severally be dwelt upon as displaying the like multiplication of effects. But it would needlessly tax the reader's patience to pursue, in their many ramifications, these various changes: here become so involved and subtle as to be followed with some difficulty.

Without further evidence, we venture to think our case is made out. The imperfections of statement which brevity has necessitated, do not, we believe, militate against the propositions laid down. The qualifications here and there demanded would not, if made, affect the inferences. Though in one instance, where sufficient evidence is not attainable, we have been unable to show that the law of Progress applies; yet there is high probability that the same generalization holds which holds throughout the rest of creation. Though, in tracing the genesis of Progress, we have frequently spoken of complex causes as if they were simple ones; it still remains true that such causes are far less complex than their results. Detailed criticisms cannot affect our main position. Endless facts go to show that every kind of progress is from the homogeneous to the heterogeneous; and that it is so because each change is followed by many changes. And it is significant that where the facts are most accessible and abundant, there are these truths most manifest.

However, to avoid committing ourselves to more than is yet proved, we must be content with saying that such are the law and the cause of all progress that is known to as. Should the Nebular Hypothesis ever be established, then it will become manifest that the Universe at large, like every organism, was once homogeneous; that as a whole, and in every detail, it has unceasingly advanced towards greater heterogeneity; and that its heterogeneity is still increasing. It will be seen that as in each event of to-day, so from the beginning, the decomposition of every expended force into several forces has been perpetually producing a higher complication; that the increase of heterogeneity so brought about is still going on, and must continue to go on; and that thus Progress is not an accident, not a thing within human control, but a beneficent necessity.

A few words must be added on the ontological bearings of our argument. Probably not a few will conclude that here is an attempted solution of the great questions with which Philosophy in all ages has perplexed itself. Let none thus deceive themselves. Only such as know not the scope and the limits of Science can fall into so grave an error. The foregoing generalizations apply, not to the genesis of things in themselves, but to their genesis as manifested to the human consciousness. After all that has been said, the ultimate mystery remains just as it was. The explanation of that which is explicable, does but bring out into greater clearness the inexplicableness of that which remains behind. However we may succeed in reducing the equation to its lowest terms, we are not thereby enabled to determine the unknown quantity: on the contrary, it only becomes more manifest that the unknown quantity can never be found.

Little as it seems to do so, fearless inquiry tends continually to give a firmer basis to all true Religion. The timid sectarian, alarmed at the progress of knowledge, obliged to abandon one by one the superstitions of his ancestors, and daily finding his cherished beliefs more and more shaken, secretly fears that all things may some day be explained; and has a corresponding dread of Science: thus evincing the profoundest of all infidelity — the fear lest the truth be bad. On the other hand, the sincere man of science, content to follow wherever the evidence leads him, becomes by each new inquiry more profoundly convinced that the Universe is an insoluble problem. Alike in the external and the internal worlds, he sees himself in the midst of perpetual changes, of which he can discover neither the beginning nor the end. If, tracing back the evolution of things, he allows himself to entertain the hypothesis that all matter once existed in a diffused form, he finds it utterly impossible to conceive how this came to be so; and equally, if he speculates on the future, he can assign no limit to the grand succession of phenomena ever unfolding themselves before him. On the other hand, if he looks inward, he perceives that both terminations of the thread of consciousness are beyond his grasp: he cannot remember when or how consciousness commenced, and he cannot examine the consciousness that at any moment exists; for only a state of consciousness that is already past can become the object of thought, and never one which is passing.

When, again, he turns from the succession of phenomena, external or internal, to their essential nature, he is equally at fault. Though he may succeed in resolving all properties of objects into manifestations of force, he is not thereby enabled to realize what force is; but finds, on the contrary, that the more he thinks about it, the more he is baffled. Similarly, though analysis of mental actions may finally bring him down to sensations as the original materials out of which all thought is woven, he is none the forwarder; for he cannot in the least comprehend sensation — cannot even conceive how sensation is possible. Inward and outward things he thus discovers to be alike inscrutable in their ultimate genesis and nature. He sees that the Materialist and Spiritualist controversy is a mere war of words; the disputants being equally absurd — each believing he understands that which it is impossible for any man to understand. In all directions his investigations eventually bring him face to face with the unknowable; and he ever more clearly perceives it to be the unknowable. He learns at once the greatness and the littleness of human intellect — its power in dealing with all that comes within the range of experience; its impotence in dealing with all that transcends experience. He feels, with a vividness which no others can, the utter incomprehensibleness of the simplest fact, considered in itself. He alone truly *sees* that absolute knowledge is impossible. He alone *knows* that under all things there lies an impenetrable mystery.

II
MANNERS AND FASHION

Whoever has studied the physiognomy of political meetings, cannot fail to have remarked a connection between democratic opinions and peculiarities of costume. At a Chartist demonstration, a lecture on Socialism, or a *soirée* of the Friends of Italy, there will be seen many among the audience, and a still larger ratio among the speakers, who get themselves up in a style more or less unusual. One gentleman on the platform divides his hair down the centre, instead, of on one side; another brushes it back off the forehead, in the fashion known as "bringing out the intellect;" a third has so long forsworn the scissors, that his locks sweep his shoulders. A considerable sprinkling of moustaches may be observed; here and there an imperial; and occasionally some courageous breaker of conventions exhibits a full-grown beard.[5] This nonconformity in hair is countenanced by various nonconformities in dress, shown by others of the assemblage. Bare necks, shirt-collars *à la* Byron, waistcoats cut Quaker fashion, wonderfully shaggy great coats, numerous oddities in form and colour, destroy the monotony usual in crowds. Even those exhibiting no conspicuous peculiarity, frequently indicate by something in the pattern or make-up of their clothes, that they pay small regard to what their tailors tell them about the prevailing taste. And when the gathering breaks up, the varieties of head gear displayed — the number of caps, and the abundance of felt hats — suffice to prove that were the world at large like-minded, the black cylinders which tyrannize over us would soon be deposed.

The foreign correspondence of our daily press shows that this relationship between political discontent and the disregard of customs exists on the Continent also. Red republicanism has always been distinguished by its hirsuteness. The authorities of Prussia, Austria, and Italy, alike recognize certain forms of hat as

---

5 This was written before moustaches and beards had become common.

indicative of disaffection, and fulminate against them accordingly. In some places the wearer of a blouse runs a risk of being classed among the *suspects*; and in others, he who would avoid the bureau of police, must beware how he goes out in any but the ordinary colours. Thus, democracy abroad, as at home, tends towards personal singularity.

Nor is this association of characteristics peculiar to modern times, or to reformers of the State. It has always existed; and it has been manifested as much in religious agitations as in political ones. Along with dissent from the chief established opinions and arrangements, there has ever been some dissent from the customary social practices. The Puritans, disapproving of the long curls of the Cavaliers, as of their principles, cut their own hair short, and so gained the name of "Roundheads." The marked religious nonconformity of the Quakers was accompanied by an equally-marked nonconformity of manners — in attire, in speech, in salutation. The early Moravians not only believed differently, but at the same time dressed differently, and lived differently, from their fellow Christians.

That the association between political independence and independence of personal conduct, is not a phenomenon of to-day only, we may see alike in the appearance of Franklin at the French court in plain clothes, and in the white hats worn by the last generation of radicals. Originality of nature is sure to show itself in more ways than one. The mention of George Fox's suit of leather, or Pestalozzi's school name, "Harry Oddity," will at once suggest the remembrance that men who have in great things diverged from the beaten track, have frequently done so in small things likewise. Minor illustrations of this truth may be gathered in almost every circle. We believe that whoever will number up his reforming and rationalist acquaintances, will find among them more than the usual proportion of those who in dress or behaviour exhibit some degree of what the world calls eccentricity.

If it be a fact that men of revolutionary aims in politics or religion, are commonly revolutionists in custom also, it is not less a fact that those whose office it is to uphold established arrangements in State and Church, are also those who most adhere to the social forms and observances bequeathed to us by past generations. Practices elsewhere extinct still linger about the headquarters of government. The monarch still gives assent to Acts of Parliament in the old French of the Normans; and Norman French terms are still used in law. Wigs, such as those we see depicted in old portraits, may yet be found on the heads of judges and barristers. The Beefeaters at the Tower wear the costume of Henry VIIth's bodyguard. The University dress of the present year varies but little from that worn soon after the Reformation. The claret-coloured coat, knee-breeches, lace shirt frills, ruffles, white silk stockings, and buckled shoes, which once formed the usual attire of a gentleman, still survive as the court-dress. And it need scarcely be said that at *levées* and drawing-rooms, the ceremonies are prescribed with an exactness, and enforced with a rigour, not else where to be found.

Can we consider these two series of coincidences as accidental and unmeaning? Must we not rather conclude that some necessary relationship obtains between them? Are there not such things as a constitutional conservatism, and a constitutional tendency to change? Is there not a class which clings to the old in all things; and another class so in love with progress as often to mistake novelty for improvement? Do we not find some men ready to bow to established authority of whatever kind; while others demand of every such authority its reason, and reject it if it fails to justify itself? And must not the minds thus contrasted tend to become respectively conformist and nonconformist, not only in politics and religion, but in other things? Submission, whether to a government, to the dogmas of ecclesiastics, or to that code of behaviour which society at large has set up, is essentially of the same nature; and the sentiment which induces resistance to the despotism of rulers, civil or spiritual, likewise induces resistance to the despotism of the world's opinion. Look at them fundamentally, and all enactment, alike of the legislature, the consistory, and the saloon — all regulations, formal or virtual, have a common character: they are all limitations of men's freedom. "Do this — Refrain from that," are the blank formulas into which they may all be written: and in each case the understanding is that obedience will bring approbation here and paradise hereafter; while disobedience will entail imprisonment, or sending to Coventry, or eternal torments, as the case may be. And if restraints, however named, and through whatever apparatus of means exercised, are one in their action upon men, it must happen that those who are patient under one kind of restraint, are likely to be patient under another; and conversely, that those impatient of restraint in general, will, on the average, tend to show their impatience in all directions.

That Law, Religion, and Manners are thus related — that their respective kinds of operation come under one generalization — that they have in certain contrasted characteristics of men a common support and a common danger — will, however, be most clearly seen on discovering that they have a common origin. Little as from present appearances we should suppose it, we shall yet find that at first, the control of religion, the control of laws, and the control of manners, were all one control. However incredible it may now seem, we believe it to be demonstrable that the rules of etiquette, the provisions of the statute-book, and the commands of the decalogue, have grown from the same root. If we go far enough back into the ages of primeval Fetishism, it becomes manifest that originally Deity, Chief, and Master of the ceremonies were identical. To make good these positions, and to show their bearing on what is to follow, it will be necessary here to traverse ground that is in part somewhat beaten, and at first sight irrelevant to our topic. We will pass over it as quickly as consists with the exigencies of the argument.

That the earliest social aggregations were ruled solely by the will of the strong man, few dispute. That from the strong man proceeded not only Monarchy, but the conception of a God, few admit: much as Carlyle and others have said in evidence of it. If, however, those who are unable to believe this, will lay aside the ideas of God and man in which they have been educated, and study the aboriginal ideas of them, they will at least see some probability in the hypothesis. Let them remember that before experience had yet taught men to distinguish between the possible and the impossible; and while they were ready on the slightest suggestion to ascribe unknown powers to any object and make a fetish of it; their conceptions of humanity and its capacities were necessarily vague, and without specific limits. The man who by unusual strength, or cunning, achieved something that others had failed to achieve, or something which they did not understand, was considered by them as differing from themselves; and, as we see in the belief of some Polynesians that only their chiefs have souls, or in that of the ancient Peruvians that their nobles were divine by birth, the ascribed difference was apt to be not one of degree only, but one of kind.

Let them remember next, how gross were the notions of God, or rather of gods, prevalent during the same era and afterwards — how concretely gods were conceived as men of specific aspects dressed in specific ways — how their names were literally "the strong," "the destroyer," "the powerful one," — how, according to the Scandinavian mythology, the "sacred duty of blood-revenge" was acted on by the gods themselves, — and how they were not only human in their vindictiveness, their cruelty, and their quarrels with each other, but were supposed to have amours on earth, and to consume the viands placed on their altars. Add to which, that in various mythologies, Greek, Scandinavian, and others, the oldest beings are giants; that according to a traditional genealogy the gods, demi-gods, and in some cases men, are descended from these after the human fashion; and that while in the East we hear of sons of God who saw the daughters of men that they were fair, the Teutonic myths tell of unions between the sons of men and the daughters of the gods.

Let them remember, too, that at first the idea of death differed widely from that, which we have; that there are still tribes who, on the decease of one of their number, attempt to make the corpse stand, and put food into his mouth; that the Peruvians had feasts at which the mummies of their dead Incas presided, when, as Prescott says, they paid attention "to these insensible remains as if they were instinct with life;" that among the Fejees it is believed that every enemy has to be killed twice; that the Eastern Pagans give extension and figure to the soul, and attribute to it all the same substances, both solid and liquid, of which our bodies are composed; and that it is the custom among most barbarous races to bury food, weapons, and trinkets along with the dead body, under the manifest belief that it will presently need them.

Lastly, let them remember that the other world, as originally conceived, is simply some distant part of this world — some Elysian fields, some happy hunting-ground, accessible even to the living, and to which, after death, men travel in anticipation of a life analogous in general character to that which they led before. Then, co-ordinating these general facts — the ascription of unknown powers to chiefs and medicine men; the belief in deities having human forms, passions, and behaviour; the imperfect comprehension of death as distinguished from life; and the proximity of the future abode to the present, both in position and character — let them reflect whether they do not almost unavoidably suggest the conclusion that the aboriginal god is the dead chief: the chief not dead in our sense, but gone away carrying with him food and weapons to some rumoured region of plenty, some promised land, whither be had long intended to lead his followers, and whence he will presently return to fetch them.

This hypothesis once entertained, is seen to harmonize with all primitive ideas and practices. The sons of the deified chief reigning after him, it necessarily happens that all early kings are held descendants of the gods; and the fact that alike in Assyria, Egypt, among the Jews, Phœnicians, and ancient Britons, kings' names were formed out of the names of the gods, is fully explained. The genesis of Polytheism out of Fetishism, by the successive migrations of the race of god-kings to the other world — a genesis illustrated in the Greek mythology, alike by the precise genealogy of the deities, and by the specifically asserted apotheosis of the later ones — tends further to bear it out. It explains the fact that in the old creeds, as in the still extant creed of the Otaheitans, every family has its guardian spirit, who is supposed to be one of their departed relatives; and that they sacrifice to these as minor gods — a practice still pursued by the Chinese and even by the Russians. It is perfectly congruous with the Grecian myths concerning the wars of the Gods with the Titans and their final usurpation; and it similarly agrees with the fact that among the Teutonic gods proper was one Freir who came among them by adoption, "but was born among the *Vanes*, a somewhat mysterious *other* dynasty of gods, who had been conquered and superseded by the stronger and more warlike Odin dynasty." It harmonizes, too, with the belief that there are different gods to different territories and nations, as there were different chiefs; that these gods contend for supremacy as chiefs do; and it gives meaning to the boast of neighbouring tribes — "Our god is greater than your god." It is confirmed by the notion universally current in early times, that the gods come from this other abode, in which they commonly live, and appear among men — speak to them, help them, punish them. And remembering this, it becomes manifest that the prayers put up by primitive peoples to their gods for aid in battle, are meant literally — that their gods are expected to come back from the other kingdom they are reigning over, and once more fight the old enemies they had before warred against so implacably; and it needs but to name the Iliad, to remind every one how thoroughly they believed the expectation fulfilled.

All government, then, being originally that of the strong man who has become a fetish by some manifestation of superiority, there arises, at his death — his supposed departure on a long projected expedition, in which he is accompanied by his slaves and concubines sacrificed at his tomb — there arises, then, the incipient division of religious from political control, of civil rule from spiritual. His son becomes deputed chief during his absence; his authority is cited as that by which his son acts; his vengeance is invoked on all who disobey his son; and his commands, as previously known or as asserted by his son, become the germ of a moral code: a fact we shall the more clearly perceive if we remember, that early moral codes inculcate mainly the virtues of the warrior, and the duty of exterminating some neighbouring tribe whose existence is an offence to the deity.

From this point onwards, these two kinds of authority, at first complicated together as those of principal and agent, become slowly more and more distinct. As experience accumulates, and ideas of causation grow more precise, kings lose their supernatural attributes; and, instead of God-king, become God-descended king, God-appointed king, the Lord's anointed, the vicegerent of heaven, ruler reigning by Divine right. The old theory, however, long clings to men in feeling, after it has disappeared in name; and "such divinity doth hedge a king," that even now, many, on first seing one, feel a secret surprise at finding him an ordinary sample of humanity. The sacredness attaching to royalty attaches afterwards to its appended institutions — to legislatures, to laws. Legal and illegal are synonymous with right and wrong; the authority of Parliament is held unlimited; and a lingering faith in governmental power continually generates unfounded hopes from its enactments. Political scepticism, however, having destroyed the divine *prestige* of royalty, goes on ever increasing, and promises ultimately to reduce the State to a purely secular institution, whose regulations are limited in their sphere, and have no other authority than the general will. Meanwhile, the religious control has been little by little separating itself from the civil, both in its essence and in its forms. While from the God-king of the savage have arisen in one direction, secular rulers who, age by age, have been losing the sacred attributes men ascribed to them; there has arisen in another direction, the conception of a deity, who, at first human in all things, has been gradually losing human materiality, human form, human passions, human modes of action: until now, anthropomorphism has become a reproach.

Along with this wide divergence in men's ideas of the divine and civil ruler has been taking place a corresponding divergence in the codes of conduct respectively proceeding from them. While the king was a deputy-god — a governor such as the Jews looked for in the Messiah — a governor considered, as the Czar still is, "our God upon Earth," — it, of course, followed that his commands were the supreme rules. But as men ceased to believe in his supernatural origin and nature, his commands ceased to be the highest; and there arose a distinction between the regulations made by him, and the regulations handed down from the old god-kings, who were rendered ever more sacred by time and the accumulation of myths. Hence came respectively, Law and Morality: the one growing ever more concrete, the other more abstract; the authority of the one ever on the decrease, that of the other ever on the increase; originally the same, but now placed daily in more marked antagonism.

Simultaneously there has been going on a separation of the institutions administering these two codes of conduct. While they were yet one, of course Church and State were one; the king was arch-priest, not nominally, but really — alike the giver of new commands and the chief interpreter of the old commands; and the deputy-priests coming out of his family were thus simply expounders of the dictates of their ancestry: at first as recollected, and afterwards as ascertained by professed interviews with them. This union — which still existed practically during the middle ages, when the authority of kings was mixed up with the authority of the pope, when there were bishop-rulers having all the powers of feudal lords, and when priests punished by penances — has been, step by step, becoming less close. Though monarchs are still "defenders of the faith," and ecclesiastical chiefs, they are but nominally such. Though bishops still have civil power, it is not what they once had. Protestantism shook loose the bonds of union; Dissent has long been busy in organizing a mechanism for the exercise of religious control, wholly independent of law; in America, a separate organization for that purpose already exists; and if anything is to be hoped from the Anti-State-Church Association — or, as it has been newly named, "The Society for the Liberation of Religion from State Patronage and Control" — we shall presently have a separate organization here also.

Thus alike in authority, in essence, and in form, political and spiritual rule have been ever more widely diverging from the same root. That increasing division of labour which marks the progress of society in other things, marks it also in this separation of government into civil and religions; and if we observe how the morality which forms the substance of religions in general, is beginning to be purified from the associated creeds, we may anticipate that this division will be ultimately carried much further.

Passing now to the third species of control — that of Manners — we shall find that this, too, while it had a common genesis with the others, has gradually come to have a distinct sphere and a special embodiment. Among early aggregations of men before yet social observances existed, the sole forms of courtesy known were the signs of submission to the strong man; as the sole law was his will and the sole religion the awe of his supposed supernaturalness. Originally, ceremonies were modes of behaviour to the god-king. Our commonest titles have been derived from his names. And all salutations were primarily worship paid to him. Let us trace out these truths in detail, beginning with titles.

The fact already noticed, that the names of early kings among divers races are formed by the addition of certain syllables to the names of their gods — which certain syllables, like our *Mac* and *Fitz*, probably mean "son of," or "descended from" — at once gives meaning to the term *Father* as a divine title. And when we read, in Selden, that "the composition out of these names of Deities was not only proper to Kings: their Grandes and more honorable Subjects" (no doubt members of the royal race) "had sometimes the like;" we see how the term *Fa-*

ther, properly used by these also, and by their multiplying descendants, came to be a title used by the people in general. And it is significant as bearing on this point, that among the most barbarous nation in Europe, where belief in the divine nature of the ruler still lingers, *Father* in this higher sense is still a regal distinction. When, again, we remember how the divinity at first ascribed to kings was not a complimentary fiction but a supposed fact; and how, further, under the Fetish philosophy the celestial bodies are believed to be personages who once lived among men; we see that the appellations of oriental rulers, "Brother to the Sun," &c., were probably once expressive of a genuine belief; and have simply, like many other things, continued in use after all meaning has gone out of them. We may infer, too, that the titles God, Lord, Divinity, were given to primitive rulers literally — that the *nostra divinitas* applied to the Roman emperors, and the various sacred designations that have been borne by monarchs, down to the still extant phrase, "Our Lord the King," are the dead and dying forms of what were once living facts. From these names, God, Father, Lord, Divinity, originally belonging to the God-king, and afterwards to God and the king, the derivation of our commonest titles of respect is clearly traceable.

There is reason to think that these titles were originally proper names. Not only do we see among the Egyptians, where Pharaoh was synonymous with king, and among the Romans, where to be Cæsar, meant to be Emperor, that the proper names of the greatest men were transferred to their successors, and so became class names; but in the Scandinavian mythology we may trace a human title of honour up to the proper name of a divine personage. In Anglo-Saxon *bealdor*, or *baldor*, means Lord; and Balder is the name of the favourite of Odin's sons — the gods who with him constitute the Teutonic Pantheon. How these names of honour became general is easily understood. The relatives of the primitive kings — the grandees described by Selden as having names formed on those of the gods, and shown by this to be members of the divine race — necessarily shared in the epithets, such as *Lord*, descriptive of superhuman relationships and nature. Their ever-multiplying offspring inheriting these, gradually rendered them comparatively common. And then they came to be applied to every man of power: partly from the fact that, in these early days when men conceived divinity simply as a stronger kind of humanity, great persons could be called by divine epithets with but little exaggeration; partly from the fact that the unusually potent were apt to be considered as unrecognized or illegitimate descendants of "the strong, the destroyer, the powerful one;" and partly, also, from compliment and the desire to propitiate.

Progressively as superstition diminished, this last became the sole cause. And if we remember that it is the nature of compliment, as we daily hear it, to attribute more than is due — that in the constantly widening application of "esquire," in the perpetual repetition of "your honour" by the fawning Irishman, and in the use of the name "gentleman" to any coalheaver or dustman by the lower classes of London, we have current examples of the depreciation of titles consequent on compliment — and that in barbarous times, when the wish to propitiate was stronger than now, this effect must have been greater; we shall see that there naturally arose an extensive misuse of all early distinctions. Hence the facts, that the Jews called Herod a god; that *Father*, in its higher sense, was a term used among them by servants to masters; that *Lord* was applicable to any person of worth and power. Hence, too, the fact that, in the later periods of the Roman Empire, every man saluted his neighbour as *Dominus* and *Rex*.

But it is in the titles of the middle ages, and in the growth of our modern ones out of them, that the process is most clearly seen. *Herr, Don, Signior, Seigneur, Sennor*, were all originally names of rulers — of feudal lords. By the complimentary use of these names to all who could, on any pretence, be supposed to merit them, and by successive degradations of them from each step in the descent to a still lower one, they have come to be common forms of address. At first the phrase in which a serf acosted his despotic chief, *mein herr* is now familiarly applied in Germany to ordinary people. The Spanish title *Don*, once proper to noblemen and gentlemen only, is now accorded to all classes. So, too, is it with *Signior* in Italy. *Seigneur*, and *Monseigneur*, by contraction in *Sieur* and *Monsieur*, have produced the term of respect claimed by every Frenchman. And whether *Sire* be or be not a like contraction of *Signior*, it is clear that, as it was borne by sundry of the ancient feudal lords of France, who, as Selden says, "affected rather to bee stiled by the name of *Sire* than

Baron, as *Le Sire de Montmorencie*, *Le Sire de Beauieu*, and the like," and as it has been commonly used to monarchs, our word *Sir*, which is derived from it, originally meant lord or king. Thus, too, is it with feminine titles. *Lady*, which, according to Horne Tooke, means *exalted*, and was at first given only to the few, is now given to all women of education. *Dame*, once an honourable name to which, in old books, we find the epithets of "high-born" and "stately" affixed, has now, by repeated widenings of its application, become relatively a term of contempt. And if we trace the compound of this, *ma Dame*, through its contractions — *Madam*, *ma'am*, *mam*, *mum*, we find that the "Yes'm" of Sally to her mistress is originally equivalent to "Yes, my exalted," or "Yes, your highness." Throughout, therefore, the genesis of words of honour has been the same. Just as with the Jews and with the Romans, has it been with the modern Europeans. Tracing these everyday names to their primitive significations of *lord* and *king*, and remembering that in aboriginal societies these were applied only to the gods and their descendants, we arrive at the conclusion that our familiar *Sir* and *Monsieur* are, in their primary and expanded meanings, terms of adoration.

Further to illustrate this gradual depreciation of titles, and to confirm the inference drawn, it may be well to notice in passing, that the oldest of them have, as might be expected, been depreciated to the greatest extent. Thus, *Master* — a word proved by its derivation and by the similarity of the connate words in other languages (Fr., *maître* for *master*; Russ., *master*; Dan., *meester*; Ger., *meister*) to have been one of the earliest in use for expressing lordship — has now become applicable to children only, and under the modification of "Mister," to persons next above the labourer. Again, knighthood, the oldest kind of dignity, is also the lowest; and Knight Bachelor, which is the lowest order of knighthood, is more ancient than any other of the orders. Similarly, too, with the peerage Baron is alike the earliest and least elevated of its divisions. This continual degradation of all names of honor has, from time to time, made it requisite to introduce new ones having that distinguishing effect which the originals had lost by generality of use; just as our habit of misapplying superlatives has, by gradually destroying their force, entailed the need for fresh ones. And if, within the last thousand years, this process has produced effects thus marked, we may readily conceive how, during previous thousands, the titles of gods and demi-gods came to be used to all persons exercising power; as they have since come to be used to persons of respectability.

If from names of honour we turn to phrases of honour, we find similar facts. The Oriental styles of address, applied to ordinary people — "I am your slave," "All I have is yours," "I am your sacrifice" — attribute to the individual spoken, to the same greatness that *Monsieur* and *My Lord* do: they ascribe to him the character of an all-powerful ruler, so immeasurably superior to the speaker as to be his owner. So, likewise, with the Polish expressions of respect — "I throw myself under your feet," "I kiss your feet." In our now meaningless subscription to a formal letter — "Your most obedient servant," — the same thing is visible. Nay, even in the familiar signature "Yours faithfully," the "yours," if interpreted as originally meant, is the expression of a slave to his master.

All these dead forms were once living embodiments of fact — were primarily the genuine indications of that submission to authority which they verbally assert; were afterwards naturally used by the weak and cowardly to propitiate those above them; gradually grew to be considered the due of such; and, by a continually wider misuse, have lost their meanings, as *Sir* and *Master* have done. That, like titles, they were in the beginning used only to the God-king, is indicated by the fact that, like titles, they were subsequently used in common to God and the king. Religious worship has ever largely consisted of professions of obedience, of being God's servants, of belonging to him to do what he will with. Like titles, therefore, these common phrases of honour had a devotional origin.

Perhaps, however, it is in the use of the word *you* as a singular pronoun that the popularizing of what were once supreme distinctions is most markedly illustrated. This speaking of a single individual in the plural, was originally an honour given only to the highest — was the reciprocal of the imperial "we" assumed by such. Yet now, by being applied to successively lower and lower classes, it has become all but universal. Only by one sect of Christians, and in a few secluded districts, is the primitive *thou* still used. And the *you*, in becoming common to all ranks has simultaneously lost every vestige of the honour once attaching to it.

But the genesis of Manners out of forms of allegiance and worship, is above all shown in men's modes of salutation. Note first the significance of the word. Among the Romans, the *salutatio* was a daily homage paid by clients and inferiors to superiors. This was alike the case with civilians and in the army. The very derivation of our word, therefore, is suggestive of submission. Passing to particular forms of obeisance (mark the word again), let us begin with the Eastern one of baring the feet. This was, primarily, a mark of reverence, alike to a god and a king. The act of Moses before the burning bush, and the practice of Mahometans, who are sworn on the Koran with their shoes off, exemplify the one employment of it; the custom of the Persians, who remove their shoes on entering the presence of their monarch, exemplifies the other. As usual, however, this homage, paid next to inferior rulers, has descended from grade to grade. In India, it is a common mark of respect; a polite man in Turkey always leaves his shoes at the door, while the lower orders of Turks never enter the presence of their superiors but in their stockings; and in Japan, this baring of the feet is an ordinary salutation of man to man.

Take another case. Selden, describing the ceremonies of the Romans, says: — "For whereas it was usual either to kiss the Images of their Gods, or adoring them, to stand somewhat off before them, solemnly moving the right hand to the lips, and then, casting it as if they had cast kisses, to turne the body on the same hand (which was the right forme of Adoration), it grew also by custom, first that the emperors, being next to Deities, and by some accounted as Deities, had the like done to them in acknowledgment of their Greatness." If, now, we call to mind the awkward salute of a village school-boy, made by putting his open hand up to his face and describing a semicircle with his forearm; and if we remember that the salute thus used as a form of reverence in country districts, is most likely a remnant of the feudal times; we shall see reason for thinking that our common wave of the hand to a friend across the street, represents what was primarily a devotional act.

Similarly have originated all forms of respect depending upon inclinations of the body. Entire prostration is the aboriginal sign of submission. The passage of Scripture, "Thou hast put all under his feet," and that other one, so suggestive in its anthropomorphism, "The Lord said unto my Lord, sit thou at my right hand, until I make thine enemies thy footstool," imply, what the Assyrian sculptures fully bear out, that it was the practice of the ancient god-kings of the East to trample upon the conquered. And when we bear in mind that there are existing savages who signify submission by placing the neck under the foot of the person submitted to, it becomes obvious that all prostration, especially when accompanied by kissing the foot, expressed a willingness to be trodden upon — was an attempt to mitigate wrath by saying, in signs, "Tread on me if you will." Remembering, further, that kissing the foot, as of the Pope and of a saint's statue, still continues it Europe to be a mark of extreme reverence; that prostration to feudal lords was once general; and that its disappearance must have taken place, not abruptly, but by gradual modification into something else; we have ground for deriving from these deepest of humiliations all inclinations of respect; especially as the transition is traceable. The reverence of a Russian serf, who bends his head to the ground, and the salaam of the Hindoo, are abridged prostrations; a bow is a short salaam; a nod is a short bow.

Should any hesitate to admit this conclusion, then perhaps, on being reminded that the lowest of these obeisances are common where the submission is most abject; that among ourselves the profundity of the bow marks the amount of respect; and lastly, that the bow is even now used devotionally in our churches — by Catholics to their altars, and by Protestants at the name of Christ — they will see sufficient evidence for thinking that this salutation also was originally worship.

The same may be said, too, of the curtsy, or courtesy, as it is otherwise written. Its derivation from *courtoisie*, courteousness, that is, behaviour like that at court, at once shows that it was primarily the reverence paid to a monarch. And if we call to mind that falling upon the knees, or upon one knee, has been a common obeisance of subjects to rulers; that in ancient manuscripts and tapestries, servants are depicted as assuming this attitude while offering the dishes to their masters at table; and that this same attitude is assumed towards our own queen at every presentation; we may infer, what the character of the curtsy itself suggests, that it is an abridged act of kneeling. As the word has been contracted from *courtoisie* into curtsy, so the motion has been contracted from a placing of the knee on the floor, to a lowering of the knee towards the floor. Moreover, when we compare the curtsy of a lady with the awkward one a peasant girl makes, which, if continued, would bring her down on both knees, we may see in this last a remnant of that greater reverence required of serfs. And when, from considering that simple kneeling of the West, still represented by the curtsy, we pass Eastward, and note the attitude of the Mahomedan worshipper, who not only kneels but bows his head to the ground, we may infer that the curtsy also, is an evanescent form of the aboriginal prostration.

In further evidence of this it may be remarked, that there has but recently disappeared from the salutations of men, an action having the same proximate derivation with the curtsy. That backward sweep of the foot with which the conventional stage-sailor accompanies his bow — a movement which prevailed generally in past generations, when "a bow and a scrape" went together, and which, within the memory of living persons, was made by boys to their schoolmaster with the effect of wearing a hole in the floor — is pretty clearly a preliminary to going on one knee. A motion so ungainly could never have been intentionally introduced; even if the artificial introduction of obeisances were possible. Hence we must regard it as the remnant of something antecedent: and that this something antecedent was humiliating may be inferred from the phrase, "scraping an acquaintance;" which, being used to denote the gaining of favour by obsequiousness, implies that the scrape was considered a mark of servility — that is, of *serf*-ility.

Consider, again, the uncovering of the head. Almost everywhere this has been a sign of reverence, alike in temples and before potentates; and it yet preserves among us some of its original meaning. Whether it rains, hails, or shines, you must keep your head bare while speaking to the monarch; and on no plea may you remain covered in a place of worship. As usual, however, this ceremony, at first a submission to gods and kings, has become in process of time a common civility. Once an acknowledgment of another's unlimited supremacy, the removal of the hat is now a salute accorded to very ordinary persons, and that uncovering, originally reserved for entrance into "the house of God," good manners now dictates on entrance into the house of a common labourer.

Standing, too, as a mark of respect, has undergone like extensions in its application. Shown, by the practice in our churches, to be intermediate between the humiliation signified by kneeling and the self-respect which sitting implies, and used at courts as a form of homage when more active demonstrations of it have been made, this posture is now employed in daily life to show consideration; as seen alike in the attitude of a servant before a master, and in that rising which politeness prescribes on the entrance of a visitor.

Many other threads of evidence might have been woven into our argument. As, for example, the significant fact, that if we trace back our still existing law of primogeniture — if we consider it as displayed by Scottish clans, in which not only ownership but government devolved from the beginning on the eldest son of the eldest — if we look further back, and observe that the old titles of lordship, *Signor, Seigneur, Sennor, Sire, Sieur*, all originally mean, senior, or elder — if we go Eastward, and find that *Sheick* has a like derivation, and that the Oriental names for priests, as *Pir*, for instance, are literally interpreted *old man* — if we note in Hebrew records how primeval is the ascribed superiority of the first-born, how great the authority of elders, and how sacred the memory of patriarchs — and if, then, we remember that among divine titles are "Ancient of Days," and "Father of Gods and men;" — we see how completely these facts harmonize with the hypothesis, that the aboriginal god is the first man sufficiently great to become a tradition, the earliest whose power and deeds made him remembered; that hence antiquity unavoidably became associated with superiority, and age with nearness in blood to "the powerful one;" that so there naturally arose that domination of the eldest which characterizes all history, and that theory of human degeneracy which even yet survives.

We might further dwell on the facts, that *Lord* signifies high-born, or, as the same root gives a word meaning heaven, possibly heaven-born; that, before it became common, *Sir* or *Sire*, as well as *Father*, was the distinction of a priest; that *worship*, originally worth-ship — a term of respect that has been used commonly, as well as to magistrates — is also our term for the act of attributing greatness or worth to the Deity; so that to ascribe worth-ship to a man is to worship him. We might make much of the evidence that all early governments are more or less distinctly theocratic; and that among ancient Eastern nations even the commonest forms and customs appear to have been influenced by religion. We might enforce our argument respecting the derivation of ceremonies, by tracing out the aboriginal obeisance made by putting dust on the head, which probably symbolizes putting the head in the dust: by affiliating the practice prevailing among certain tribes, of doing another honour by presenting him with a portion of hair torn from the head — an act which seems tantamount to saying, "I am your slave;" by investigating the Oriental custom of giving to a visitor any object he speaks of admiringly, which is pretty clearly a carrying out the compliment, "All I have is yours."

Without enlarging, however, on these and many minor facts, we venture to think that the evidence already assigned is sufficient to justify our position. Had the proofs been few or of one kind, little faith could have been placed in the inference. But numerous as they are, alike in the case of titles, in that of complimentary phrases, and in that of salutes — similar and simultaneous, too, as the process of depreciation has been in all of these; the evidences become strong by mutual confirmation. And when we recollect, also, that not only have the results of this process been visible in various nations and in all times, but that they are occurring among ourselves at the present moment, and that the causes assigned for previous depreciations may be seen daily working out other ones — when we recollect this, it becomes scarcely possible to doubt that the process has been as alleged; and that our ordinary words, acts, and phrases of civility were originally acknowledgments of submission to another's omnipotence.

Thus the general doctrine, that all kinds of government exercised over men were at first one government — that the political, the religious, and the ceremonial forms of control are divergent brandies of a general and once indivisible control — begins to look tenable. When, with the above facts fresh in mind, we read primitive records, and find that "there were giants in those days" — when we remember that in Eastern traditions Nimrod, among others, figures in all the characters of giant, king, and divinity — when we turn to the sculptures exhumed by Mr. Layard, and contemplating in them the effigies of kings driving over enemies, trampling on prisoners, and adored by prostrate slaves, then observe how their actions correspond to the primitive names for the divinity, "the strong," "the destroyer," "the powerful one" — when we find that the earliest temples were also the residences of the kings — and when, lastly, we discover that among races of men still living, there are current superstitions analogous to those which old records and old buildings indicate; we begin to realize the probability of the hypothesis that has been set forth.

Going back, in imagination, to the remote era when men's theories of things were yet unformed; and conceiving to ourselves the conquering chief as dimly figured in ancient myths, and poems, and ruins; we may see that all rules of conduct whatever spring from his will. Alike legislator and judge, all quarrels among his subjects are decided by him; and his words become the Law. Awe of him is the incipient Religion; and his maxims furnish its first precepts. Submission is made to him in the forms he prescribes; and these give birth to Manners. From the first, time developes political allegiance and the administration of justice; from the second, the worship of a being whose personality becomes ever more vague, and the inculcation of precepts ever more abstract; from the third, forms of honour and the rules of etiquette.

In conformity with the law of evolution of all organized bodies, that general functions are gradually separated into the special functions constituting them, there have grown up in the social organism for the better performance of the governmental office, an apparatus of law-courts, judges, and barristers; a national church, with its bishops and priests; and a system of caste, titles, and ceremonies, administered by society at large. By the first, overt aggressions are cognized and punished; by the second, the disposition to commit such aggressions is in some degree checked; by the third, those minor breaches of good conduct, which the others do not notice, are denounced and chastised. Law and Religion control behaviour in its essentials: Manners control it in its details. For regulating those daily actions which are too numerous and too unimportant to be officially directed, there comes into play this subtler set of restraints. And when we consider what these restraints are — when we analyze the words, and phrases, and salutes employed we see that in origin as in effect, the system is a setting up of temporary governments between all men who come in contact, for the purpose of better managing the intercourse between them.

From the proposition, that these several kinds of government are essentially one, both in genesis and function, may be deduced several important corollaries, directly bearing on our special topic.

Let us first notice, that there is not only a common origin and office for all forms of rule, but a common necessity for them. The aboriginal man, coming fresh from the killing of bears and from lying in ambush for his enemy, has, by the necessities of his condition, a nature requiring to be curbed in its every impulse. Alike in war and in the chase, his daily discipline has been that of sacrificing other creatures to his own needs and passions. His character, bequeathed to him by ancestors who led similar lives, is moulded by this discipline — is fitted to this existence. The unlimited selfishness, the love of inflicting pain, the bloodthirstiness, thus kept active, he brings with him into the social state. These dispositions put him in constant danger of conflict with his equally savage neighbour. In small things as in great, in words as in deeds, he is aggressive; and is hourly liable to the aggressions of others like natured. Only, therefore, by the most rigorous control exercised over all actions, can the primitive unions of men be maintained. There must be a ruler strong, remorseless, and of indomitable will; there must be a creed terrible in its threats to the disobedient; and there must be the most servile submission of all inferiors to superiors. The law must be cruel; the religion must be stern; the ceremonies must be strict.

The co-ordinate necessity for these several kinds of restraint might be largely illustrated from history were there space. Suffice it to point out, that where the civil power has been weak, the multiplication of thieves, assassins, and banditti, has indicated the approach of social dissolution; that when, from the corruptness of its ministry, religion has lost its influence, as it did just before the Flagellants appeared, the State has been endangered; and that the disregard of established social observances has ever been an accompaniment of political revolutions. Whoever doubts the necessity for a government of manners proportionate in strength to the co-existing political and religious governments, will be convinced on calling to mind that until recently even elaborate codes of behaviour failed to keep gentlemen from quarrelling in the streets and fighting duels in taverns; and on remembering further, that even now people exhibit at the doors of a theatre, where there is no ceremonial law to rule them, a degree of aggressiveness which would produce confusion if carried into social intercourse.

As might be expected, we find that, having a common origin and like general functions, these several controlling agencies act during each era with similar degrees of vigour. Under the Chinese despotism, stringent and multitudinous in its edicts and harsh in the enforcement of them, and associated with which there is an equally stem domestic despotism exercised by the eldest surviving male of the family, there exists a system of observances alike complicated and rigid. There is a tribunal of ceremonies. Previous to presentation at court, ambassadors pass many days in practising the required forms. Social intercourse is cumbered by endless compliments and obeisances. Class distinctions are strongly marked by badges. The chief regret on losing an only son is, that there will be no one to perform the sepulchral rites. And if there wants a definite measure of the respect paid to social ordinances, we have it in the torture to which ladies submit in having their feet crushed. In India, and indeed throughout the East, there exists a like connection between the pitiless tyranny of rulers, the dread terrors of immemorial creeds, and the rigid restraint of unchangeable customs: the caste regulations continue still unalterable; the fashions of clothes and furniture have remained the same for ages; suttees are so ancient as to be mentioned by Strabo and Diodorus Siculus; justice is still administered at the palace-gates as of old; in short, "every usage is a precept of religion and a maxim of jurisprudence."

A similar relationship of phenomena was exhibited in Europe during the Middle Ages. While all its governments were autocratic, while feudalism held sway, while the Church was unshorn of its power, while the criminal code was full of horrors and the hell of the popular creed full of terrors, the rules of behaviour were both more numerous and more carefully conformed to than now. Differences of dress marked divisions of rank. Men were limited by law to a certain width of shoe-toes; and no one below a specified degree might wear a cloak less than so many inches long. The symbols on banners and shields were carefully attended to. Heraldry was an important branch of knowledge. Precedence was strictly insisted on. And those various salutes of which we now use the abridgments were gone through in full. Even during our own last century, with its corrupt House of Commons and little-curbed monarchs, we may mark a correspondence of social formalities. Gentlemen were still distinguished from lower classes by dress; people sacrificed themselves to inconvenient requirements — as powder, hooped petticoats, and towering head-dresses; and children addressed their parents as *Sir* and *Madam*.

A further corollary naturally following this last, and almost, indeed, forming part of it, is, that these several kinds of government decrease in stringency at the same rate. Simultaneously with the decline in the influence of priesthoods, and in the fear of eternal torments — simultaneously with the mitigation of political tyranny, the growth of popular power, and the amelioration of criminal codes; has taken place that diminution of formalities and that fading of distinctive marks, now so observable. Looking at home, we may note that there is less attention to precedence than there used to be. No one in our day ends an interview with the phrase "your humble servant." The employment of the word *Sir*, once general in social intercourse, is at present considered bad breeding; and on the occasions calling for them, it is held vulgar to use the words "Your Majesty," or "Your Royal Highness," more than once in a conversation. People no longer formally drink each other's healths; and even the taking wine with each other at dinner has ceased to be fashionable. The taking-off of hats between gentlemen has been gradually falling into disuse. Even when the hat is removed, it is no longer swept out at arm's length, but is simply lifted. Hence the remark made upon us by foreigners, that we take off our hats less than any other nation in Europe — a remark that should be coupled with the other, that we are the freest nation in Europe.

As already implied, this association of facts is not accidental. These titles of address and modes of salutation, bearing about them, as they all do, something of that servility which marks their origin, become distasteful in proportion as men become more independent themselves, and sympathise more with the independence of others. The feeling which makes the modern gentleman tell the labourer standing bareheaded before him to put on his hat — the feeling which gives us a dislike to those who cringe and fawn — the feeling which makes us alike assert our own dignity and respect that of others — the feeling which thus leads us more and more to discountenance all forms and names which confess inferiority and submission; is the same feeling which resists despotic power and inaugurates popular government, denies the authority of the Church and establishes the right of private judgment.

A fourth fact, akin to the foregoing, is, that these several kinds of government not only decline together, but corrupt together. By the same process that a Court of Chancery becomes a place not for the administration of justice, but for the withholding of it — by the same process that a national church, from being an agency for moral control, comes to be merely a thing of formulas and tithes and bishoprics — by this same process do titles and ceremonies that once had a meaning and a power become empty forms.

Coats of arms which served to distinguish men in battle, now figure on the carriage panels of retired grocers. Once a badge of high military rank, the shoulder-knot has become, on the modern footman, a mark of servitude. The name Banneret, which once marked a partially-created Baron — a Baron who had passed his military "little go" — is now, under the modification of Baronet, applicable to any one favoured by wealth or interest or party feeling. Knighthood has so far ceased to be an honour, that men now honour themselves by declining it. The military dignity *Escuyer* has, in the modern Esquire, become a wholly unmilitary affix. Not only do titles, and phrases, and salutes cease to fulfil their original functions, but the whole apparatus of social forms tends to become useless for its original purpose — the facilitation of social intercourse. Those most learned in ceremonies, and most precise in the observance of them, are not always the best behaved; as those deepest read in creeds and scriptures are not therefore the most religious; nor those who have the clearest notions of legality and illegality, the most honest. Just as lawyers are of all men the least noted for probity; as cathedral towns have a lower moral character than most others; so, if Swift is to be believed, courtiers are "the most insignificant race of people that the island can afford, and with the smallest tincture of good manners."

But perhaps it is in that class of social observances comprehended under the term Fashion, which we must here discuss parenthetically, that this process of corruption is seen with the greatest distinctness. As contrasted with Manners, which dictate our minor acts in relation to other persons, Fashion dictates our minor acts in relation to ourselves. While the one prescribes that part of our deportment which directly affects our neighbours; the other prescribes that part of our deportment which is primarily personal, and in which our neighbours are concerned only as spectators. Thus distinguished as they are, however, the two have a common source. For while, as we have shown, Manners originate by imitation of the behaviour pursued *towards* the great; Fashion originates by imitation *of* the behaviour of the great. While the one has its derivation in the titles, phrases, and salutes used *to* those in power; the other is derived from the habits and appearances exhibited *by* those in power.

The Carrib mother who squeezes her child's head into a shape like that of the chief; the young savage who makes marks on himself similar to the scars carried by the warriors of his tribe (which is probably the origin of tattooing); the Highlander who adopts the plaid worn by the head of his clan; the courtiers who affect greyness, or limp, or cover their necks, in imitation of their king; and the people who ape the courtiers; are alike acting under a kind of government connate with that of Manners, and, like it too, primarily beneficial. For notwithstanding the numberless absurdities into which this copyism has led the people, from nose-rings to ear-rings, from painted faces to beauty-spots, from shaven heads to powdered wigs, from filed teeth and stained nails to bell-girdles, peaked shoes and breeches stuffed with bran, — it must yet be concluded, that as the strong men, the successful men, the men of will, intelligence, and originality, who have got to the top, are, on the average, more likely to show judgment in their habits and tastes than the mass, the imitation of such is advantageous.

By and by, however, Fashion, corrupting like these other forms of rule, almost wholly ceases to be an imitation of the best, and becomes an imitation of quite other than the best. As those who take orders are not those having a special fitness for the priestly office, but those who see their way to a living by it; as legislators and public functionaries do not become such by virtue of their political insight and power to rule, but by virtue of birth, acreage, and class influence; so, the self-elected clique who set the fashion, gain this prerogative, not by their force of nature, their intellect, their higher worth or better taste, but gain it solely by their unchecked assumption. Among the initiated are to be found neither the noblest in rank, the chief in power, the best cultured, the most refined, nor those of greatest genius, wit, or beauty; and their reunions, so far from being superior to others, are noted for their inanity. Yet, by the example of these sham great, and not by that of the truly great, does society at large now regulate its goings and comings, its hours, its dress, its small usages. As a natural consequence, these have generally little or none of that suitableness which the theory of fashion implies they should have. But instead of a continual progress towards greater elegance and convenience, which might be expected to occur did people copy the ways of the really best, or follow their own ideas of propriety, we have a reign of mere whim, of unreason, of change for the sake of change, of wanton oscillations from either extreme to the other — a reign of usages without meaning, times without fitness, dress without taste. And thus life *à la mode*, instead of being life conducted in the most rational manner, is life regulated by spendthrifts and idlers, milliners and tailors, dandies and silly women.

To these several corollaries — that the various orders of control exercised over men have a common origin and a common function, are called out by co-ordinate necessities and co-exist in like stringency, decline together and corrupt together — it now only remains to add that they become needless together. Consequent as all kinds of government are upon the unfitness of the aboriginal man for social life; and diminishing in coerciveness as they all do in proportion as this unfitness diminishes; they must one and all come to an end as humanity acquires complete adaptation to its new conditions. That discipline of circumstances which has already wrought out such great changes in us, must go on eventually to work out yet greater ones. That daily curbing of the lower nature and culture of the higher, which out of cannibals and devil worshippers has evolved philanthropists, lovers of peace, and haters of superstition, cannot fail to evolve out of these, men as much superior to them as they are to their progenitors. The causes that have produced past modifications are still in action; must continue in action as long as there exists any incongruity between man's desires and the requirements of the social state; and must eventually make him organically fit for the social state. As it is now needless to forbid man-eating and Fetishism, so will it ultimately become needless to forbid murder, theft, and the minor offences of our criminal code. When human nature has grown into conformity with the moral law, there will need no judges and statute-books; when it spontaneously takes the right course in all things, as in some things it does already, prospects of future reward or punishment will not be wanted as incentives; and when fit behaviour has become instinctive, there will need no code of ceremonies to say how behaviour shall be regulated.

Thus, then, may be recognised the meaning, the naturalness, the necessity of those various eccentricities of reformers which we set out by describing. They are not accidental; they are not mere personal caprices, as people are apt to suppose. On the contrary, they are inevitable results of the law of relationship above illustrated. That community of genesis, function, and decay which all forms of restraint exhibit, is simply the obverse of the fact at first pointed out, that they have in two sentiments of human nature a common preserver and a common destroyer. Awe of power originates and cherishes them all: love of freedom undermines and periodically weakens them all. The one defends despotism and asserts the supremacy of laws, adheres to old creeds and supports ecclesiastical authority, pays respect to titles and conserves forms; the other, putting rectitude above legality, achieves periodical instalments of political liberty, inaugurates Protestantism and works out its consequences, ignores the senseless dictates of Fashion and emancipates men from dead customs.

To the true reformer no institution is sacred, no belief above criticism. Everything shall conform itself to equity and reason; nothing shall be saved by its prestige. Conceding to each man liberty to pursue his own ends and satisfy his own tastes, he demands for himself like liberty; and consents to no restrictions on this, save those which other men's equal claims involve. No matter whether it be an ordinance of one man, or an ordinance of all men, if it trenches on his legitimate sphere of action, he denies its validity. The tyranny that would impose on him a particular style of dress and a set mode of behaviour, he resists equally with the tyranny that would limit his buyings and sellings, or dictate his creed. Whether the regulation be formally made by a legislature, or informally made by society at large — whether the penalty for disobedience be imprisonment, or frowns and social ostracism, he sees to be a question of no moment. He will utter his belief notwithstanding the threatened punishment; he will break conventions spite of the petty persecutions that will be visited on him. Show him that his actions are inimical to his fellow-men, and he will pause. Prove that he is disregarding their legitimate claims — that he is doing what in the nature of things must produce unhappiness; and he will alter his course. But until you do this — until you demonstrate that his proceedings are essentially inconvenient or inelegant, essentially irrational, unjust, or ungenerous, he will persevere.

Some, indeed, argue that his conduct *is* unjust and ungenerous. They say that he has no right to annoy other people by his whims; that the gentleman to whom his letter comes with no "Esq." appended to the address, and the lady whose evening party he enters with gloveless hands, are vexed at what they consider his want of respect, or want of breeding; that thus his eccentricities cannot be indulged save at the expense of his neighbours' feelings; and that hence his nonconformity is in plain terms selfishness.

He answers that this position, if logically developed, would deprive men of all liberty whatever. Each must conform all his acts to the public taste, and not his own. The public taste on every point having been once ascertained, men's habits must thenceforth remain for ever fixed; seeing that no man can adopt other habits without sinning against the public taste, and giving people disagreeable feelings. Consequently, be it an era of pig-tails or high-heeled shoes, of starched ruffs or trunk-hose, all must continue to wear pig-tails, high-heeled shoes, starched ruffs, or trunk-hose to the crack of doom.

If it be still urged that he is not justified in breaking through others' forms that he may establish his own, and so sacrificing the wishes of many to the wishes of one, he replies that all religious and political changes might be negatived on like grounds. He asks whether Luther's sayings and doings were not extremely offensive to the mass of his contemporaries; whether the resistance of Hampden was not disgusting to the time-servers around him; whether every reformer has not shocked men's prejudices, and given immense displeasure by the opinions he uttered. The affirmative answer he follows up by demanding what right the reformer has, then, to utter these opinions; whether he is not sacrificing the feelings of many to the feelings of one: and so proves that, to be consistent, his antagonists must condemn not only all nonconformity in actions, but all nonconformity in thoughts.

His antagonists rejoin that *his* position, too, may be pushed to an absurdity. They argue that if a man may offend by the disregard of some forms, he may as legitimately do so by the disregard of all; and they inquire — Why should he not go out to dinner in a dirty shirt, and with an unshorn chin? Why should he not spit on the drawing-room carpet, and stretch his heels up to the mantel-shelf?

The convention-breaker answers, that to ask tins, implies a confounding of two widely-different classes of actions — the actions that are *essentially* displeasurable to those around, with the actions that are but *incidentally* displeasurable to them. He whose skin is so unclean as to offend the nostrils of his neighbours, or he who talks so loudly as to disturb a whole room, may be justly complained of, and rightly excluded by society from its assemblies. But he who presents himself in a surtout in place of a dress-coat, or in brown trousers instead of black, gives offence not to men's senses, or their innate tastes, but merely to their prejudices, their bigotry of convention. It cannot be said that his costume is less elegant or less intrinsically appropriate than the one prescribed; seeing that a few hours earlier in the day it is admired. It is the implied rebellion, therefore, that annoys. How little the cause of quarrel has to do with the dress itself, is seen in the fact that a century ago black clothes would have been thought preposterous for hours of recreation, and that a few years hence some now forbidden style may be nearer the requirements of Fashion than the present one. Thus the reformer explains that it is not against the natural restraints, but against the artificial ones, that he protests; and that manifestly the fire of sneers and angry glances which he has to bear, is poured upon him because he will not bow down to the idol which society has set up.

Should he be asked how we are to distinguish between conduct that is *absolutely* disagreeable to others, and conduct that is *relatively* so, he answers, that they will distinguish themselves, if men will let them. Actions intrinsically repugnant will ever be frowned upon, and must ever remain as exceptional as now. Actions not intrinsically repugnant will establish themselves as proper. No relaxation of customs will introduce the practice of going to a party in muddy boots, and with unwashed hands; for the dislike of dirt would continue were Fashion abolished tomorrow. That love of approbation which now makes people so solicitous to be *en règle* would still exist — would still make them careful of their personal appearance — would still induce them to seek admiration by making themselves ornamental — would still cause them to respect the natural laws of good behaviour, as they now do the artificial ones. The change would simply be from a repulsive monotony to a picturesque variety. And if there be any regulations respecting which it is uncertain whether they are based on reality or on convention, experiment will soon decide, if due scope be allowed.

When at length the controversy comes round, as controversies often do, to the point whence it started, and the "party of order" repeat their charge against the rebel, that he is sacrificing the feelings of others to the gratification of his own wilfulness, he replies once for all that they cheat themselves by misstatements. He accuses them of being so despotic, that, not content with being masters over their own ways and habits, they would be masters over his also; and grumble because he will not let them. He merely asks the same freedom which they exercise; they, however, propose to regulate his course as well as their own — to cut and clip his mode of life into agreement with their approved pattern; and then charge him with wilfulness and selfishness, because he does not quietly submit! He warns them that he shall resist, nevertheless; and that he shall do so, not only for the assertion of his own independence, but for their good. He tells them that they are slaves, and know it not; that they are shackled, and kiss their chains; that they have lived all their days in prison, and complain at the walls being broken down. He says he mast persevere, however, with a view to his own release; and in spite of their present expostulations, he prophesies that when they have recovered from the fright which the prospect of freedom produces, they will thank him for aiding in their emancipation.

Unamiable as seems this find-fault mood, offensive as is this defiant attitude, we must beware of overlooking the truths enunciated, in dislike of the advocacy. It is an unfortunate hindrance to all innovation, that in virtue of their very function, the innovators stand in a position of antagonism; and the disagreeable manners, and sayings, and doings, which this antagonism generates, are commonly associated with the doctrines promulgated. Quite forgetting that whether the thing attacked be good or bad, the combative spirit is necessarily repulsive; and quite forgetting that the toleration of abuses seems amiable merely from its passivity; the mass of men contract a bias against advanced views, and in favour of stationary ones, from intercourse with their respective adherents. "Conservatism," as Emerson says, "is debonnair and social; reform is individual and imperious." And this remains true, however vicious the system conserved, however righteous the reform to be effected. Nay, the indignation of the purists is usually extreme in proportion as the evils to be got rid of are great. The more urgent the required change, the more intemperate is the vehemence of its promoters. Let no one, then, confound with the principles of this social nonconformity the acerbity and the disagreeable self-assertion of those who first display it.

The most plausible objection raised against resistance to conventions, is grounded on its impolicy, considered even from the progressist's point of view. It is urged by many of the more liberal and intelligent — usually those who have themselves shown some independence of behaviour in earlier days — that to rebel in these small matters is to destroy your own power of helping on reform in greater matters. "If you show yourself eccentric in manners or dress, the world," they say, "will not listen to you. You will be considered as crotchety, and impracticable. The opinions you express on important subjects, which might have been treated with respect had you conformed on minor points, will now inevitably be put down among your singularities; and thus, by dissenting in trifles, you disable yourself from spreading dissent in essentials."

Only noting, as we pass, that this is one of those anticipations which bring about their own fulfilment — that it is because most who disapprove these conventions do not show their disapproval, that the few who do show it look eccentric — and that did all act out their convictions, no such inference as the above would be drawn, and no such evil would result; — noting this as we pass, we go on to reply that these social restraints, and forms, and requirements, are not small evils, but among the greatest. Estimate their sum total, and we doubt whether they would not exceed most others. Could we add up the trouble, the cost, the jealousies, vexations, misunderstandings, the loss of time and the loss of pleasure, which these conventions entail — could we clearly realize the extent to which we are all daily hampered by them, daily enslaved by them; we should perhaps come to the conclusion that the tyranny of Mrs. Grundy is worse than any other tyranny we suffer under. Let us look at a few of its hurtful results; beginning with those of minor importance.

It produces extravagance. The desire to be *comme il faut*, which underlies all conformities, whether of manners, dress, or styles of entertainment, is the desire which makes many a spendthrift and many a bankrupt. To "keep up appearances," to have a house in an approved quarter furnished in the latest taste, to give expensive dinners and crowded *soirées*, is an ambition forming the natural outcome of the conformist spirit. It is needless to enlarge on these follies: they have been satirized by hosts of writers, and in every drawing-room. All that here concerns us, is to point out that the respect for social observances, which men think so praiseworthy, has the same root with this effort to be fashionable in mode of living; and that, other things equal, the last cannot be diminished without the first being diminished also. If, now, we consider all that this extravagance entails — if we count up the robbed tradesmen, the stinted governesses, the ill-educated children, the fleeced relatives, who have to suffer from it — if we mark the anxiety and the many moral delinquencies which its perpetrators involve themselves in; we shall see that this regard for conventions is not quite so innocent as it looks.

Again, it decreases the amount of social intercourse. Passing over the reckless, and those who make a great display on speculation with the occasional result of getting on in the world to the exclusion of much better men, we come to the far larger class who, being prudent and honest enough not to exceed their means, and yet having a strong wish to be "respectable," are obliged to limit their entertainments to the smallest possible number; and that each of these may be turned to the greatest advantage in meeting the claims upon their hospitality, are induced to issue their invitations with little or no regard to the comfort or mutual fitness of their guests. A few inconveniently-large assemblies, made up of people mostly strange to each other or but distantly acquainted, and having scarcely any tastes in common, are made to serve in place of many small parties of friends intimate enough to have some bond of thought and sympathy. Thus the quantity of intercourse is diminished, and the quality deteriorated. Because it is the custom to make costly preparations and provide costly refreshments; and because it entails both less expense and less trouble to do this for many persons on a few occasions than for few persons on many occasions; the reunions of our less wealthy classes are rendered alike infrequent and tedious.

Let it be further observed, that the existing formalities of social intercourse drive away many who most need its refining influence: and drive them into injurious habits and associations. Not a few men, and not the least sensible men either, give up in disgust this going out to stately dinners, and stiff evening-parties; and instead, seek society in clubs, and cigar-divans, and taverns. "I'm sick of this standing about in drawing-rooms, talking nonsense, and trying to look happy," will answer one of them when taxed with his desertion. "Why should I any longer waste time and money, and temper? Once I was ready enough to rush home from the office to dress; I sported embroidered shirts, submitted to tight boots, and cared nothing for tailors' and haberdashers' bills. I know better now. My patience lasted a good while; for though I found each night pass stupidly, I always hoped the next would make amends. But I'm undeceived. Cab-hire and kid gloves cost more than any evening party pays for; or rather — it is worth the cost of them to avoid the party. No, no; I'll no more of it. Why should I pay five shillings a time for the privilege of being bored?"

If, now, we consider that this very common mood tends towards billiard-rooms, towards long sittings over cigars and brandy-and-water, towards Evans's and the Coal Hole, towards every place where amusement may be had; it becomes a question whether these precise observances which hamper our set meetings, have not to answer for much of the prevalent dissoluteness. Men must have excitements of some kind or other; and if debarred from higher ones will fall back upon lower. It is not that those who thus take to irregular habits are essentially those of low tastes. Often it is quite the reverse. Among half a dozen intimate friends, abandoning formalities and sitting at ease round the fire, none will enter with greater enjoyment into the highest kind of social intercourse — the genuine communion of thought and feeling; and if the circle includes women of intelligence and refinement, so much the greater is their pleasure. It is because they will no longer be choked with the mere dry husks of conversation which society offers them, that they fly its assemblies, and seek those with whom they may have discourse that is at least real, though unpolished. The men who thus long for substantial mental sympathy, and will go where they can get it, are often, indeed, much better at the core than the men who are content with the inanities of gloved and scented party-goers — men who feel no need to come morally nearer to their fellow creatures than they can come while standing, tea-cup in hand, answering trifles with trifles; and who, by feeling no such need, prove themselves shallow-thoughted and cold-hearted.

It is true, that some who shun drawing-rooms do so from inability to bear the restraints prescribed by a genuine refinement, and that they would be greatly improved by being kept under these restraints. But it is not less true that, by adding to the legitimate restraints, which are based on convenience and a regard for others, a host of factitious restraints based only on convention, the refining discipline, which would else have been borne with benefit, is rendered unbearable, and so misses its end. Excess of government invariably defeats itself by driving away those to be governed. And if over all who desert its entertainments in disgust either at their emptiness or their formality, society thus loses its salutary influence — if such not only fail to receive that moral culture which the company of ladies, when rationally regulated, would give them, but, in default of other relaxation, are driven into habits and companionships which often end in gambling and drunkenness; must we not say that here, too, is an evil not to be passed over as insignificant?

Then consider what a blighting effect these multitudinous preparations and ceremonies have upon the pleasures they profess to subserve. Who, on calling to mind the occasions of his highest social enjoyments, does not find them to have been wholly informal, perhaps impromptu? How delightful a picnic of friends, who forget all observances save those dictated by good nature! How pleasant the little unpretended gatherings of book-societies, and the like; or those purely accidental meetings of a few people well known to each other! Then, indeed, we may see that "a man sharpeneth the countenance of his friend." Cheeks flush, and eyes sparkle. The witty grow brilliant, and even the dull are excited into saying good things. There is an overflow of topics; and the right thought, and the right wards to put it in, spring up unsought. Grave alternates with gay: now serious converse, and now jokes, anecdotes, and playful raillery. Everyone's best nature is shown everyone's best feelings are in pleasurable activity; and, for the time, life seems well worth having.

Go now and dress for some half-past eight dinner, or some ten o'clock "at home;" and present yourself in spotless attire, with every hair arranged to perfection. How great the difference! The enjoyment seems in the inverse ratio of the preparation. These figures, got up with such finish and precision, appear but half alive. They have frozen each other by their primness; and your faculties feel the numbing effects of the atmosphere the moment you enter it. All those thoughts, so nimble and so apt awhile since, have disappeared — have suddenly acquired a preternatural power of eluding you. If you venture a remark to your neighbour, there comes a trite rejoinder, and there it ends. No subject you can hit upon outlives half a dozen sentences. Nothing that is said excites any real interest in you; and you feel that all you say is listened to with apathy. By some strange magic, things that usually give pleasure seem to have lost all charm.

You have a taste for art. Weary of frivolous talk, you turn to the table, and find that the book of engravings and the portfolio of photographs are as flat as the conversation. You are fond of music. Yet the singing, good as it is, you hear with utter indifference; and say "Thank you" with a tense of being a profound hypocrite. Wholly at ease though you could be, for your own part, you find that your sympathies will not let you. You see young gentlemen feeling whether their ties are properly adjusted, looking vacantly round, and considering what they shall do next. You see ladies sitting disconsolately, waiting for some one to speak to them, and wishing they had the wherewith to occupy their fingers. You see the hostess standing about the doorway, keeping a factitious smile on her face, and racking her brain to find the requisite nothings with which to greet her guests as they enter. You see numberless traits of weariness and embarrassment; and, if you have any fellow feeling, these cannot fail to produce a feeling of discomfort. The disorder is catching; and do what you will you cannot resist the general infection. You struggle against it; you make spasmodic efforts to be lively; but none of your sallies or your good stories do more than raise a simper or a forced laugh: intellect and feeling are alike asphyxiated. And when, at length, yielding to your disgust, you rush away, how great is the relief when you get into the fresh air, and see the stars! How you "Thank God, that's over!" and half resolve to avoid all such boredom for the future!

What, now, is the secret of this perpetual miscarriage and disappointment? Does not the fault lie with all these needless adjuncts — these elaborate dressings, these set forms, these expensive preparations, these many devices and arrangements that imply trouble and raise expectation? Who that has lived thirty years in the world has not discovered that Pleasure is coy; and must not be too directly pursued, but must be caught unawares? An air from a street-piano, heard while at work, will often gratify more than the choicest music played at a concert by the most accomplished musicians. A single good picture seen in a dealer's window, may give keener enjoyment than a whole exhibition gone through with catalogue and pencil. By the time we have got ready our elaborate apparatus by which to secure happiness, the happiness is gone. It is too subtle to be contained in these receivers, garnished with compliments, and fenced round with etiquette. The more we multiply and complicate appliances, the more certain are we to drive it away.

The reason is patent enough. These higher emotions to which social intercourse ministers, are of extremely complex nature; they consequently depend for their production upon very numerous conditions; the more numerous the conditions, the greater the liability that one or other of them will be disturbed, and the emotions consequently prevented. It takes a considerable misfortune to destroy appetite; but cordial sympathy with those around may be extinguished by a look or a word. Hence it follows, that the more multiplied the *unnecessary* requirements with which social intercourse is surrounded, the less likely are its pleasures to be achieved. It is difficult enough to fulfil continuously all the *essentials* to a pleasurable communion with others: how much more difficult, then, must it be continuously to fulfil a host of *non-essentials* also! It is, indeed, impossible. The attempt inevitably ends in the sacrifice of the first to the last — the essentials to the non-essentials. What chance is there of getting any genuine response from the lady who is thinking of your stupidity in taking her in to dinner on the wrong arm? How are you likely to have agreeable converse with the gentleman who is fuming internally because he is not placed next to the hostess? Formalities, familiar as they may become, necessarily occupy attention — necessarily multiply the occasions for mistake, misunderstanding, and jealousy, on the part of one or other — necessarily distract all minds from the thoughts and feelings that should occupy them — necessarily, therefore, subvert those conditions under which only any sterling intercourse is to be had.

And this indeed is the fatal mischief which these conventions entail — a mischief to which every other is secondary. They destroy those highest of our pleasures which they profess to subserve. All institutions are alike in this, that however useful, and needful even, they originally were, they not only in the end cease to be so, but become detrimental. While humanity is growing, they continue fixed; daily get more mechanical and unvital; and by and by tend to strangle what they before preserved. It is not simply that they become corrupt and fail to act they become obstructions. Old forms of government finally grow so oppressive, that they must be thrown off even at the risk of reigns of terror. Old creeds end in being dead formulas, which no longer aid but distort and arrest the general mind; while the State-churches administering them, come to be instruments for subsidizing conservatism and repressing progress. Old schemes of education, incarnated in public schools and colleges, continue filling the heads of new generations with what has become relatively useless knowledge, and, by consequence, excluding knowledge which is useful. Not an organization of any kind — political, religious, literary, philanthropic — but what, by its ever-multiplying regulations, its accumulating wealth, its yearly addition of officers, and the creeping into it of patronage and party feeling, eventually loses its original spirit, and sinks into a mere lifeless mechanism, worked with a view to private ends — a mechanism which not merely fails of its first purpose, but is a positive hindrance to it.

Thus is it, too, with social usages. We read of the Chinese that they have "ponderous ceremonies transmitted from time immemorial," which make social intercourse a burden. The court forms prescribed by monarchs for their own exaltation, have, in all times and places, ended in consuming the comfort of their lives. And so the artificial observances of the dining-room and saloon, in proportion as they are many and strict, extinguish that agreeable communion which they were originally intended to secure. The dislike with which people commonly speak of society that is "formal," and "stiff," and "ceremonious," implies the general recognition of this fact; and this recognition, logically developed, involves that all usages of behaviour which are not based on natural requirements, are injurious. That these conventions defeat their own ends is no new assertion. Swift, criticising the manners of his day, says — "Wise men are often more uneasy at the over-civility of these refiners than they could possibly be in the conversation of peasants and mechanics."

But it is not only in these details that the self-defeating action of our arrangements is traceable: it is traceable in the very substance and nature of them. Our social intercourse, as commonly managed, is a mere semblance of the reality sought. What is it that we want? Some sympathetic converse with our fellow-creatures: some converse that shall not be mere dead words, but the vehicle of living thoughts and feelings — converse in which the eyes and the face shall speak, and the tones of the voice be full of meaning — converse which shall make us feel no longer alone, but shall draw us closer to another, and double our own emotions by adding another's to them. Who is there that has not, from time to time, felt how cold and flat is all this talk about politics and science, and the new books and the new men, and how a genuine utterance of fellow-feeling outweighs the whole of it? Mark the words of Bacon: — "For a crowd is not a company, and faces are but a gallery of pictures, and talk but a tinkling cymbal, where there is no love."

If this be true, then it is only after acquaintance has grown into intimacy, and intimacy has ripened into friendship, that the real communion which men need becomes possible. A rationally-formed circle must consist almost wholly of those on terms of familiarity and regard, with but one or two strangers. What folly, then, underlies the whole system of our grand dinners, our "at homes," our evening parties — assemblages made up of many who never met before, many others who just bow to each other, many others who though familiar feel mutual indifference, with just a few real friends lost in the general mass! You need but look round at the artificial expressions of face, to see at once how it is. All have their disguises on; and how can there be sympathy between masks? No wonder that in private every one exclaims against the stupidity of these gatherings. No wonder that hostesses get them up rather because they must than because they wish. No wonder that the invited go less from the expectation of pleasure than from fear of giving offence. The whole thing is a gigantic mistake — an organized disappointment.

And then note, lastly, that in this case, as in all others, when an organization has become effete and inoperative for its legitimate purpose, it is employed for quite other ones — quite opposite ones. What is the usual plea put in for giving and attending these tedious assemblies? "I admit that they are stupid and frivolous enough," replies every man to your criticisms; "but then, you know, one must keep up one's connections." And could you get from his wife a sincere answer, it would be — "Like you, I am sick of these frivolities; but then, we must get our daughters married." The one knows that there is a profession to push, a practice to gain, a business to extend: or parliamentary influence, or county patronage, or votes, or office, to be got: position, berths, favours, profit. The other's thoughts runs upon husbands and settlements, wives and dowries. Worthless for their ostensible purpose of daily bringing human beings into pleasurable relations with each other, these cumbrous appliances of our social intercourse are now perseveringly kept in action with a view to the pecuniary and matrimonial results which they indirectly produce.

Who then shall say that the reform of our system of observances is unimportant? When we see how this system induces fashionable extravagance, with its entailed bankruptcy and ruin — when we mark how greatly it limits the amount of social intercourse among the less wealthy classes — when we find that many who most need to be disciplined by mixing with the refined are driven away by it, and led into dangerous and often fatal courses — when we count up the many minor evils it inflicts, the extra work which its costliness entails on all professional and mercantile men, the damage to public taste in dress and decoration by the setting up of its absurdities as standards for imitation, the injury to health indicated in the faces of its devotees at the close of the London season, the mortality of milliners and the like, which its sudden exigencies yearly involve; — and when to all these we add its fatal sin, that it blights, withers up, and kills, that high enjoyment it professedly ministers to — that enjoyment which is a chief end of our hard struggling in life to obtain — shall we not conclude that to reform our system of etiquette and fashion, is an aim yielding to few in urgency?

There needs, then, a protestantism in social usages. Forms that have ceased to facilitate and have become obstructive — whether political, religious, or other — have ever to be swept away; and eventually are so swept away in all cases. Signs are not wanting that some change is at hand. A host of satirists, led on by Thackeray, have been for years engaged in bringing our sham-festivities, and. our fashionable follies, into contempt; and in their candid moods, most men laugh at the frivolities with which they and the world in general are deluded. Ridicule has always been a revolutionary agent. That which is habitually assailed with sneers and sarcasms cannot long survive. Institutions that have lost their roots in men's respect and faith are doomed; and the day of their dissolution is not far off. The time is approaching, then, when our system of social observances must pass through some crisis, out of which it will come purified and comparatively simple.

How this crisis will be brought about, no one can with any certainty say. Whether by the continuance and increase of individual protests, or whether by the union of many persons for the practice and propagation of some better system, the future alone can decide. The influence of dissentients acting without cooperation, seems, under the present state of things, inadequate. Standing severally alone, and having no well-defined views; frowned on by conformists, and expostulated with even by those who secretly sympathize with them; subject to petty persecutions, and unable to trace any benefit produced by their example;

they are apt, one by one, to give up their attempts as hopeless. The young convention-breaker eventually finds that he pays too heavily for his nonconformity. Hating, for example, everything that bears about it any remnant of servility, he determines, in the ardour of his independence, that he will uncover to no one. But what he means simply as a general protest, he finds that ladies interpret into a personal disrespect. Though he sees that, from the days of chivalry downwards, these marks of supreme consideration paid to the other sex have been but a hypocritical counterpart to the actual subjection in which men have held them — a pretended submission to compensate for a real domination; and though he sees that when the true dignity of women is recognised, the mock dignities given to them will be abolished; yet he does not like to be thus misunderstood, and so hesitates in his practice.

In other cases, again, his courage fails him. Such of his unconventionalities as can be attributed only to eccentricity, he has no qualms about: for, on the whole, he feels rather complimented than otherwise in being considered a disregarder of public opinion. But when they are liable to be put down to ignorance, to ill-breeding, or to poverty, he becomes a coward. However clearly the recent innovation of eating some kinds of fish with knife and fork proves the fork-and-bread practice to have had little but caprice for its basis, yet he dares not wholly ignore that practice while fashion partially maintains it. Though he thinks that a silk handkerchief is quite as appropriate for drawing-room use as a white cambric one, he is not altogether at ease in acting out his opinion. Then, too, he begins to perceive that his resistance to prescription brings round disadvantageous results which he had not calculated upon. He had expected that it would save him from a great deal of social intercourse of a frivolous kind — that it would offend the fools, but not the sensible people; and so would serve as a self-acting test by which those worth knowing would be separated from those not worth knowing. But the fools prove to be so greatly in the majority that, by offending them, he closes against himself nearly all the avenues though which the sensible people are to be reached. Thus he finds, that his nonconformity is frequently misinterpreted; that there are but few directions in which he dares to carry it consistently out; that the annoyances and disadvantages which it brings upon him are greater than he anticipated; and that the chances of his doing any good are very remote. Hence he gradually loses resolution, and lapses, step by step, into the ordinary routine of observances.

Abortive as individual protests thus generally turn out, it may possibly be that nothing effectual will be done until there arises some organized resistance to this invisible despotism, by which our modes and habits are dictated. It may happen, that the government of Manners and Fashion will be rendered less tyrannical, as the political and religious governments have been, by some antagonistic union. Alike in Church and State, men's first emancipations from excess of restriction were achieved by numbers, bound together by a common creed or a common political faith. What remained undone while there were but individual schismatics or rebels, was effected when there came to be many acting in concert. It is tolerably clear that these earliest instalments of freedom could not have been obtained in any other way; for so long as the feeling of personal independence was weak and the rule strong, there could never have been a sufficient number of separate dissentients to produce the desired results. Only in these later times, during which the secular and spiritual controls have been growing less coercive, and the tendency towards individual liberty greater, has it become possible for smaller and smaller sects and parties to fight against established creeds and laws; until now men may safely stand even alone in their antagonism.

The failure of individual nonconformity to customs, as above illustrated, suggests that an analogous series of changes may have to be gone through in this case also. It is true that the *lex non scripta* differs from the *lex scripta* in this, that, being unwritten, it is more readily altered; and that it has, from time to time, been quietly ameliorated. Nevertheless, we shall find that the analogy holds substantially good. For in this case, as in the others, the essential revolution is not the substituting of any one set of restraints for any other, but the limiting or abolishing the authority which prescribes restraints. Just as the fundamental change inaugurated by the Reformation, was not a superseding of one creed by another, but an ignoring of the arbiter who before dictated creeds — just as the fundamental change which Democracy long ago commenced, was not from this particular law to that, but from the despotism of one to the freedom of all;

so, the paralled change yet to be wrought out in this supplementary government of which we are treating, is not the replacing of absurd usages by sensible ones, but the dethronement of that secret, irresponsible power which now imposes our usages, and the assertion of the right of all individuals to choose their own usages. In rules of living, a West-end clique is our Pope; and we are all papists, with but a mere sprinkling of heretics. On all who decisively rebel, comes down the penalty of excommunication, with its long catalogue of disagreeable and, indeed, serious consequences.

The liberty of the subject asserted in our constitution, and ever on the increase, has yet to be wrested from this subtler tyranny. The right of private judgment, which our ancestors wrung from the church, remains to be claimed from this dictator of our habits. Or, as before said, to free us from these idolatries and superstitious conformities, there has still to come a protestantism in social usages. Parallel, therefore, as is the change to be wrought out, it seems not improbable that it may be wrought out in an analogous way. That influence which solitary dissentients fail to gain, and that perseverance which they lack, may come into existence when they unite. That persecution which the world now visits upon them from mistaking their nonconformity for ignorance or disrespect, may diminish when it is seen to result from principle. The penalty which exclusion now entails may disappear when they become numerous enough to form visiting circles of their own. And when a successful stand has been made, and the brunt of the opposition has passed, that large amount of secret dislike to our observances which now pervades society, may manifest itself with sufficient power to effect the desired emancipation.

Whether such will be the process, time alone can decide. That community of origin, growth, supremacy, and decadence, which we have found among all kinds of government, suggests a community in modes of change also. On the other hand, Nature often performs substantially similar operations, in ways apparently different. Hence these details can never be foretold.

Meanwhile, let us glance at the conclusions that have been reached. On the one side, government, originally one, and afterwards subdivided for the better fulfilment of its function, must be considered as having ever been, in all its branches — political, religious, and ceremonial — beneficial; and, indeed, absolutely necessary. On the other side, government, under all its forms, must be regarded as subserving a temporary office, made needful by the unfitness of aboriginal humanity for social life; and the successive diminutions of its coerciveness in State, in Church, and in Custom, must be looked upon as steps towards its final disappearance. To complete the conception, there requires to be borne in mind the third fact, that the genesis, the maintenance, and the decline of all governments, however named, are alike brought about by the humanity to be controlled: from which may be drawn the inference that, on the average, restrictions of every kind cannot last much longer than they are wanted, and cannot be destroyed much faster than they ought to be.

Society, in all its developments, undergoes the process of exuviation. These old forms which it successively throws off, have all been once vitally united with it — have severally served as the protective envelopes within which a higher humanity was being evolved. They are cast aside only when they become hindrances — only when some inner and better envelope has been formed; and they bequeath to us all that there was in them good. The periodical abolitions of tyrannical laws have left the administration of justice not only uninjured, but purified. Dead and buried creeds have not carried with them the essential morality they contained, which still exists, uncontaminated by the sloughs of superstition. And all that there is of justice and kindness and beauty, embodied in our cumbrous forms of etiquette, will live perennially when the forms themselves have been forgotten.

III
THE GENESIS OF SCIENCE

There has ever prevailed among men a vague notion that scientific knowledge differs in nature from ordinary knowledge. By the Greeks, with whom Mathematics — literally *things learnt* — was alone considered as knowledge proper, the distinction must have been strongly felt; and it has ever since maintained itself in the general mind. Though, considering the contrast between the achievements of science and those of daily unmethodic thinking, it is not surprising that such a distinction has been assumed; yet it needs but to rise a little above the common point of view, to see that no such distinction can really exist: or that at best, it is but a superficial distinction. The same faculties are employed in both cases; and in both cases their mode of operation is fundamentally the same.

If we say that science is organized knowledge, we are met by the truth that all knowledge is organized in a greater or less degree — that the commonest actions of the household and the field presuppose facts colligated, inferences drawn, results expected; and that the general success of these actions proves the data by which they were guided to have been correctly put together. If, again, we say that science is prevision — is a seeing beforehand — is a knowing in what times, places, combinations, or sequences, specified phenomena will be found; we are yet obliged to confess that the definition includes much that is utterly foreign to science in its ordinary acceptation. For example, a child's knowledge of an apple. This, as far as it goes consists in previsions. When a child sees a certain form and colours, it knows that if it puts out its hand it will have certain impressions of resistance, and roundness, and smoothness; and if it bites, a certain taste. And manifestly its general acquaintance with surrounding objects is of like nature — is made up of facts concerning them, so grouped as that any part of a group being perceived, the existence of the other facts included in it is foreseen.

If, once more, we say that science is *exact* prevision, we still fail to establish the supposed difference. Not only do we find that much of what we call science is not exact, and that some of it, as physiology, can never become exact; but we find further, that many of the previsions constituting the common stock alike of wise and ignorant, *are* exact. That an unsupported body will fall; that a lighted candle will go out when immersed in water; that ice will melt when thrown on the fire — these, and many like predictions relating to the familiar properties of things have as high a degree of accuracy as predictions are capable of. It is true that the results predicated are of a very general character; but it is none the less true that they are rigorously correct as far as they go: and this is all that is requisite to fulfil the definition. There is perfect accordance between the anticipated phenomena and the actual ones; and no more than this can be said of the highest achievements of the sciences specially characterised as exact.

Seeing thus that the assumed distinction between scientific knowledge and common knowledge is not logically justifiable; and yet feeling, as we must, that however impossible it may be to draw a line between them, the two are not practically identical; there arises the question — What is the relationship that exists between them? A partial answer to this question may be drawn from the illustrations just given. On reconsidering them, it will be observed that those portions of ordinary knowledge which are identical in character with scientific knowledge, comprehend only such combinations of phenomena as are directly cognizable by the senses, and are of simple, invariable nature. That the smoke from a fire which she is lighting will ascend, and that the fire will presently boil water, are previsions which the servant-girl makes equally well with the most learned physicist; they are equally certain, equally exact with his; but they are previsions concerning phenomena in constant and direct relation — phenomena that follow visibly and immediately after their antecedents — phenomena of which the causation is neither remote nor obscure — phenomena which may be predicted by the simplest possible act of reasoning.

If, now, we pass to the previsions constituting what is commonly known as science — that an eclipse of the moon will happen at a specified time; and when a barometer is taken to the top of a mountain of known height, the mercurial column will descend a stated number of inches; that the poles of a galvanic battery immersed in water will give off, the one an inflammable and the other an inflaming gas, in definite ratio — we perceive that the relations involved are not of a kind habitually presented to our senses; that they depend, some of them, upon special combinations of causes; and that in some of them the connection between antecedents and consequents is established only by an elaborate series of inferences. The broad distinction, therefore, between the two orders of knowledge, is not in their nature, but in their remoteness from perception.

If we regard the cases in their most general aspect, we see that the labourer, who, on hearing certain notes in the adjacent hedge, can describe the particular form and colours of the bird making them; and the astronomer, who, having calculated a transit of Venus, can delineate the black spot entering on the sun's disc, as it will appear through the telescope, at a specified hour; do essentially the same thing. Each knows that on fulfilling the requisite conditions, he shall have a preconceived impression — that after a definite series of actions will come a group of sensations of a foreknown kind. The difference, then, is not in the fundamental character of the mental acts; or in the correctness of the previsions accomplished by them; but in the complexity of the processes required to achieve the previsions. Much of our commonest knowledge is, as far as it goes, rigorously precise. Science does not increase this precision; cannot transcend it. What then does it do? It reduces other knowledge to the same degree of precision. That certainty which direct perception gives us respecting coexistences and sequences of the simplest and most accessible kind, science gives us respecting coexistences and sequences, complex in their dependencies or inaccessible to immediate observation. In brief, regarded from this point of view, science may be called *an extension of the perceptions by means of reasoning.*

On farther considering the matter, however, it will perhaps be felt that this definition does not express the whole fact — that inseparable as science may be from common knowledge, and completely as we may fill up the gap between the simplest previsions of the child and the most recondite ones of the natural philosopher, by interposing a series of previsions in which the complexity of reasoning involved is greater and greater, there is yet a difference between the two beyond that which is here described. And this is true. But the difference is still not such as enables us to draw the assumed line of demarcation. It is a difference not between common knowledge and scientific knowledge; but between the successive phases of science itself, or knowledge itself — whichever we choose to call it. In its earlier phases science attains only to *certainty* of foreknowledge; in its later phases it further attains to *completeness*. We begin by discovering *a* relation: we end by discovering *the* relation. Our first achievement is to foretell the *kind* of phenomenon which will occur under specific conditions: our last achievement is to foretell not only the kind but the *amount*. Or, to reduce the proposition to its most definite form — undeveloped science is *qualitative* prevision: developed science is *quantitative* prevision.

This will at once be perceived to express the remaining distinction between the lower and the higher stages of positive knowledge. The prediction that a piece of lead will take more force to lift it than a piece of wood of equal size, exhibits certainty, but not completeness, of foresight. The kind of effect in which the one body will exceed the other is foreseen; but not the amount by which it will exceed. There is qualitative prevision only. On the other hand, the prediction that at a stated time two particular planets will be in conjunction; that by means of a lever having arms in a given ratio, a known force will raise just so many pounds; that to decompose a specified quantity of sulphate of iron by carbonate of soda will require so many grains — these predictions exhibit foreknowledge, not only of the nature of the effects to be produced, but of the magnitude, either of the effects themselves, of the agencies producing them, or of the distance in time or space at which they will be produced. There is not only qualitative but quantitative prevision.

And this is the unexpressed difference which leads us to consider certain orders of knowledge as especially scientific when contrasted with knowledge in general. Are the phenomena *measurable?* is the test which we unconsciously employ. Space is measurable: hence Geometry. Force and space are measurable: hence Statics. Time, force, and space are measurable: hence Dynamics. The invention of the barometer enabled men to extend the principles of mechanics to the atmosphere; and Aerostatics existed. When a thermometer was devised there arose a science of heat, which was before impossible. Such of our sensations as we have not yet found modes of measuring do not originate sciences. We have no science of smells; nor have we one of tastes. We have a science of the relations of sounds differing in pitch, because we have discovered a way to measure them; but we have no science of sounds in respect to their loudness or their *timbre*, because we have got no measures of loudness and *timbre*.

Obviously it is this reduction of the sensible phenomena it represents, to relations of magnitude, which gives to any division of knowledge its especially scientific character. Originally men's knowledge of weights and forces was in the same condition as their knowledge of smells and tastes is now — a knowledge not extending beyond that given by the unaided sensations; and it remained so until weighing instruments and dynamometers were invented. Before there were hour-glasses and clepsydras, most phenomena could be estimated as to their durations and intervals, with no greater precision than degrees of hardness can be estimated by the fingers. Until a thermometric scale was contrived, men's judgments respecting relative amounts of heat stood on the same footing with their present judgments respecting relative amounts of sound. And as in these initial stages, with no aids to observation, only the roughest comparisons of cases could be made, and only the most marked differences perceived; it is obvious that only the most simple laws of dependence could be ascertained — only those laws which being uncomplicated with others, and not disturbed in their manifestations, required no niceties of observation to disentangle them. Whence it appears not only that in proportion as knowledge becomes quantitative do its previsions become complete as well as certain, but that until its assumption of a quantitative character it is necessarily confined to the most elementary relations.

Moreover it is to be remarked that while, on the one hand, we can discover the laws of the greater proportion of phenomena only by investigating them quantitatively; on the other hand we can extend the range of our quantitative previsions only as fast as we detect the laws of the results we predict. For clearly the ability to specify the magnitude of a result inaccessible to direct measurement, implies knowledge of its mode of dependence on something which can be measured — implies that we know the particular fact dealt with to be an instance of some more general fact. Thus the extent to which our quantitative previsions have been carried in any direction, indicates the depth to which our knowledge reaches in that direction. And here, as another aspect of the same fact, we may further observe that as we pass from qualitative to quantitative prevision, we pass from inductive science to deductive science. Science while purely inductive is purely qualitative: when inaccurately quantitative it usually consists of part induction, part deduction: and it becomes accurately quantitative only when wholly deductive. We do not mean that the deductive and the quantitative are coextensive; for there is manifestly much deduction that is qualitative only. We mean that all quantitative prevision is reached deductively; and that induction can achieve only qualitative prevision.

Still, however, it must not be supposed that these distinctions enable us to separate ordinary knowledge from science; much as they seem to do so. While they show in what consists the broad contrast between the extreme forms of the two, they yet lead us to recognise their essential identity; and once more prove the difference to be one of degree only. For, on the one hand, the commonest positive knowledge is to some extent quantitative; seeing that the amount of the foreseen result is known within certain wide limits. And, on the other hand, the highest quantitative prevision does not reach the exact truth, but only a very near approximation to it. Without clocks the savage knows that the day is longer in the summer than in the winter; without scales he knows that stone is heavier than flesh: that is, he can foresee respecting certain results that their amounts will exceed these, and be less than those — he knows *about* what they will be. And, with his most delicate instruments and most elaborate calculations, all that the man of science can do, is to reduce the difference between the foreseen and the actual results to an unimportant quantity.

Moreover, it must be borne in mind not only that all the sciences are qualitative in their first stages, — not only that some of them, as Chemistry, have but recently reached the quantitative stage — but that the most advanced sciences have attained to their present power of determining quantities not present to the senses, or not directly measurable, by a slow process of improvement extending through thousands of years. So that science and the knowledge of the uncultured are alike in the nature of their previsions, widely as they differ in range; they possess a common imperfection, though this is immensely greater in the last than in the first; and the transition from the one to the other has been through a series of steps by which the imperfection has been rendered continually less, and the range continually wider.

These facts, that science and the positive knowledge of the uncultured cannot be separated in nature, and that the one is but a perfected and extended form of the other, must necessarily underlie the whole theory of science, its progress, and the relations of its parts to each other. There must be serious incompleteness in any history of the sciences, which, leaving out of view the first steps of their genesis, commences with them only when they assume definite forms. There must be grave defects, if not a general untruth, in a philosophy of the sciences considered in their interdependence and development, which neglects the inquiry how they came to be distinct sciences, and how they were severally evolved out of the chaos of primitive ideas.

Not only a direct consideration of the matter, but all analogy, goes to show that in the earlier and simpler stages must be sought the key to all subsequent intricacies. The time was when the anatomy and physiology of the human being were studied by themselves — when the adult man was analyzed and the relations of parts and of functions investigated, without reference either to the relations exhibited in the embryo or to the homologous relations existing in other creatures. Now, however, it has become manifest that no true conceptions, no true generalizations, are possible under such conditions. Anatomists and physiologists now find that the real natures of organs and tissues can be ascertained only by tracing their early evolution; and that the affinities between existing genera can be satisfactorily made out only by examining the fossil genera to which they are allied. Well, is it not clear that the like must be true concerning all things that undergo development? Is not science a growth? Has not science, too, its embryology? And must not the neglect of its embryology lead to a misunderstanding of the principles of its evolution and of its existing organization?

There are *à priori* reasons, therefore, for doubting the truth of all philosophies of the sciences which tacitly proceed upon the common notion that scientific knowledge and ordinary knowledge are separate; instead of commencing, as they should, by affiliating the one upon the other, and showing how it gradually came to be distinguishable from the other. We may expect to find then generalizations essentially artificial; and we shall not be deceived. Some illustrations of this may here be fitly introduced, by way of preliminary to a brief sketch of the genesis of science from the point of view indicated. And we cannot more readily find such illustrations than by glancing at a few of the various *classifications* of the sciences that have from time to time been proposed. To consider all of them would take too much space: we must content ourselves with some of the latest.

Commencing with those which may be soonest disposed of, let us notice first the arrangement propounded by Oken. An abstract of it runs thus: —

Part I. Mathesis. — *Pneumatogeny*: Primary Art, Primary Consciousness, God, Primary Rest, Time, Polarity, Motion, Man, Space, Point, Line, Surface, Globe, Rotation. — *Hylogeny*: Gravity, Matter, Ether, Heavenly Bodies, Light, Heat, Fire.

(He explains that Mathesis is the doctrine of the whole; *Pneumatogeny* being the doctrine of immaterial totalities, and *Hylogeny* that of material totalities.)

Part II. Ontology. — *Cosmogeny*: Rest, Centre, Motion, Line, Planets, Form, Planetary System, Comets. — *Stöchiogeny*: Condensation, Simple Matter, Elements, Air, Water, Earth. — *Stöchiology*: Functions of the Elements, &c. &c. — *Kingdoms of Nature*: Individuals.

(He says in explanation that "Ontology teaches us the phenomena of matter. The first of these are the heavenly bodies comprehended by *Cosmogeny*. These divide into elements — *Stöchiogeny*. The earth element divides into minerals — *Mineralogy*. These unite into one collective body — *Geogeny*. The whole in singulars is the living, or *Organic*, which again divides into plants and animals. *Biology*, therefore, divides into *Organogeny, Phytosophy, Zoosophy.*")

First Kingdom. — Minerals. *Mineralogy, Geology*.

Part III. Biology. — *Organosophy, Phytogeny, Phytophysiology, Phytology, Zoogeny, Physiology, Zoology, Psychology.'*

A glance over this confused scheme shows that it is an attempt to classify knowledge, not after the order in which it has been, or may be, built up in the human consciousness; but after an assumed order of creation. It is a pseudo-scientific cosmogony, akin to those which men have enunciated from the earliest times downwards; and only a little more respectable. As such it will not be thought worthy of much consideration by those who, like ourselves, hold that experience is the sole origin of knowledge. Otherwise, it might have been needful to dwell on the incongruities of the arrangements — to ask how motion can be treated of before space? how there can be rotation without matter to rotate? how polarity can be dealt with without involving points and lines? But it will serve our present purpose just to point out a few of the extreme absurdities resulting from the doctrine which Oken seems to hold in common with Hegel, that "to philosophize on Nature is to re-think the great thought of Creation." Here is a sample: —

"Mathematics is the universal science; so also is Physio-philosophy, although it is only a part, or rather but a condition of the universe; both are one, or mutually congruent.

"Mathematics is, however, a science of mere forms without substance. Physio-philosophy is, therefore, *mathematics endowed with substance*."

From the English point of view it is sufficiently amusing to find such a dogma not only gravely stated, but stated as an unquestionable truth. Here we see the experiences of quantitative relations which men have gathered from surrounding bodies and generalized (experiences which had been scarcely at all generalized at the beginning of the historic period) — we find these generalized experiences, these intellectual abstractions, elevated into concrete actualities, projected back into Nature, and considered as the internal frame-work of things — the skeleton by which matter is sustained. But this new form of the old realism, is by no means the most startling of the physio-philosophic principles. We presently read that,

"The highest mathematical idea, or the fundamental principle of all mathematics is the zero = 0."

"Zero is in itself nothing. Mathematics is based upon nothing, and, *consequently*, arises out of nothing.

"Out of nothing, *therefore*, it is possible for something to arise; for mathematics, consisting of propositions, is something, in relation to 0."

By such "consequentlys" and "therefores" it is, that men philosophize when they "re-think the great thought of creation." By dogmas that pretend to be reasons, nothing is made to generate mathematics; and by clothing mathematics with matter, we have the universe! If now we deny, as we *do* deny, that the highest mathematical idea is the zero; — if, on the other hand, we assert, as we *do* assert, that the fundamental idea underlying all mathematics, is that of equality; the whole of Oken's cosmogony disappears. And here, indeed, we may see illustrated, the distinctive peculiarity of the German method of procedure in these matters — the bastard *à priori* method, as it may be termed. The legitimate *à priori* method sets out with propositions of which the negation is inconceivable; the *à priori* method as illegitimately applied, sets out either with propositions of which the negation is not inconceivable, or with propositions like Oken's, of which the *affirmation* is inconceivable.

It is needless to proceed further with the analysis; else might we detail the steps by which Oken arrives at the conclusions that "the planets are coagulated colours, for they are coagulated light; that the sphere is the expanded nothing;" that gravity is "a weighty nothing, a heavy essence, striving towards a centre;" that "the earth is the identical, water the indifferent, air the different; or the first the centre, the second the radius, the last the periphery of the general globe or of fire." To comment on them would be nearly as absurd as are the propositions themselves. Let us pass on to another of the German systems of knowledge — that of Hegel.

The simple fact that Hegel puts Jacob Bœhme on a par with Bacon, suffices alone to show that his stand-point is far remote from the one usually regarded as scientific: so far remote, indeed, that it is not easy to find any common basis on which to found a criticism. Those who hold that the mind is moulded into conformity with surrounding things by the agency of surrounding things, are necessarily at a loss how to deal with those, who, like Schelling and Hegel, assert that surrounding things are solidified mind — that Nature is "petrified intelligence." However, let us briefly glance at Hegel's classification. He divides philosophy into three parts: —

1. *Logic*, or the science of the idea in itself, the pure idea.

2. *The Philosophy of Nature*, or the science of the idea considered under its other form — of the idea as Nature.

3. *The Philosophy of the Mind*, or the science of the idea in its return to itself.

Of these, the second is divided into the natural sciences, commonly so called; so that in its more detailed form the series runs thus: — Logic, Mechanics, Physics, Organic Physics, Psychology.

Now, if we believe with Hegel, first, that thought is the true essence of man; second, that thought is the essence of the world; and that, therefore, there is nothing but thought; his classification, beginning with the science of pure thought, may be acceptable. But otherwise, it is an obvious objection to his arrangement, that thought implies things thought of — that there can be no logical forms without the substance of experience — that the science of ideas and the science of things must have a simultaneous origin. Hegel, however, anticipates this objection, and, in his obstinate idealism, replies, that the contrary is true; that all contained in the forms, to become something, requires to be thought: and that logical forms are the foundations of all things.

It is not surprising that, starting from such premises, and reasoning after this fashion, Hegel finds his way to strange conclusions. Out of *space* and *time* he proceeds to build up *motion, matter, repulsion, attraction, weight*, and *inertia*. He then goes on to logically evolve the solar system. In doing this he widely diverges from. the Newtonian theory; reaches by syllogism the conviction that the planets are the most perfect celestial bodies; and, not being able to bring the stars within his theory, says that they are mere formal existences and not living matter, and that as compared with the solar system they are as little admirable as a cutaneous eruption or a swarm of flies.[6]

Results so outrageous might be left as self-disproved, were it not that speculators of this class are not alarmed by any amount of incongruity with established beliefs. The only efficient mode of treating systems like this of Hegel, is to show that

---

6 It is somewhat curious that the author of "The Plurality of Worlds," with quite other aims, should have persuaded himself into similar conclusions.

they are self-destructive — that by their first steps they ignore that authority on which all their subsequent steps depend. If Hegel professes, as he manifestly does, to develop his scheme by reasoning — if he presents successive inferences as *necessarily following* from certain premises; he implies the postulate that a belief winch necessarily follows after certain antecedents is a true belief: and, did an opponent reply to one of his inferences, that, though it was impossible to think the opposite, yet the opposite was true, he would consider the reply irrational. The procedure, however, which he would thus condemn as destructive of all thinking whatever, is just the procedure exhibited in the enunciation of his own first principles.

Mankind find themselves unable to conceive that there can be thought without things thought of. Hegel, however, asserts that there *can* be thought without things thought of. That ultimate test of a true proposition — the inability of the human mind to conceive the negation of it — which in all other cases he considers valid, he considers invalid where it suits his convenience to do so; and yet at the same time denies the right of an opponent to follow his example. If it is competent for him to posit dogmas, which are the direct negations of what human consciousness recognises; then is it also competent for his antagonists to stop him at every step in his argument by saying, that though the particular inference he is drawing seems to his mind, and to all minds, necessarily to follow from the premises, yet it is not true, but the contrary inference is true. Or, to state the dilemma in another form: — If he sets out with inconceivable propositions, then may he with equal propriety make all his succeeding propositions inconceivable ones — may at every step throughout his reasoning draw exactly the opposite conclusion to that which seems involved.

Hegel's mode of procedure being thus essentially suicidal, the Hegelian classification which depends upon it, falls to the ground. Let us consider next that of M. Comte.

As all his readers must admit, M. Comte presents us with a scheme of the sciences which, unlike the foregoing ones, demands respectful consideration. Widely as we differ from him, we cheerfully bear witness to the largeness of his views, the clearness of his reasoning, and the value of his speculations as contributing to intellectual progress. Did we believe a serial arrangement of the sciences to be possible, that of M. Comte would certainly be the one we should adopt. His fundamental propositions are thoroughly intelligible; and if not true, have a great semblance of truth. His successive steps are logically co-ordinated; and he supports his conclusions by a considerable amount of evidence — evidence which, so long as it is not critically examined, or not met by counter evidence, seems to substantiate his positions. But it only needs to assume that antagonistic attitude which *ought* to be assumed towards new doctrines, in the belief that, if true, they will prosper by conquering objectors — it needs but to test his leading doctrines either by other facts than those he cites, or by his own facts differently applied, to at once show that they will not stand. We will proceed thus to deal with the general principle on which he bases his hierarchy of the sciences.

In the second chapter of his *Cours de I'hilosophie Positive*, M. Comte says: — "Our problem is, then, to find the one *rational* order, amongst a host of possible systems."..."This order is determined by the degree of simplicity, or, what comes to the same thing, of generality of their phenomena." And the arrangement he deduces runs thus: *Mathematics, Astronomy, Physics, Chemistry, Physiology, Social Physics*. This he asserts to be "the true *filiation* of the sciences." He asserts further, that the principle of progression from a greater to a less degree of generality, "which gives this order to the whole body of science, arranges the parts of each science." And, finally, he asserts that the gradations thus established *à priori* among the sciences, and the parts of each science, "is in essential conformity with the order which has spontaneously taken place among the branches of natural philosophy;" or, in other words — corresponds with the order of historic development.

Let us compare these assertions with the facts. That there may be perfect fairness, let us make no choice, but take as the field for our comparison, the succeeding section treating of the first science — Mathematics; and let us use none but M. Comte's own facts, and his own admissions. Confining ourselves to this one science, of course our comparisons must be between its several parts. M. Comte says, that the parts of each science must be arranged in the order of their decreasing generality; and that this order of decreasing generality agrees with the order of historic development. Our inquiry must be, then, whether the history of mathematics confirms this statement.

Carrying out his principle, M. Comte divides Mathematics into "Abstract Mathematics, or the Calculus (taking the word in its most extended sense) and Concrete Mathematics, which is composed of General Geometry and of Rational Mechanics." The subject-matter of the first of these is *number*; the subject-matter of the second includes *space, time, motion, force*. The one possesses the highest possible degree of generality; for all things whatever admit of enumeration. The others are less general; seeing that there are endless phenomena that are not cognizable either by general geometry or rational mechanics. In conformity with the alleged law, therefore, the evolution of the calculus must throughout have preceded the evolution of the concrete sub-sciences. Now somewhat awkwardly for him, the first remark M. Comte makes bearing upon this point is, that "from an historical point of view, mathematical analysis *appears to have risen out of* the contemplation of geometrical and mechanical facts." True, he goes on to say that, "it is not the less independent of these sciences logically speaking;" for that "analytical ideas are, above all others, universal, abstract, and simple and geometrical conceptions are necessarily founded on them."

We will not take advantage of this last passage to charge M. Comte with teaching, after the fashion of Hegel, that there can be thought without things thought of. We are content simply to compare the two assertions, that analysis arose out of the contemplation of geometrical and mechanical facts, and that geometrical conceptions are founded upon analytical ones. Literally interpreted, they exactly cancel each other. Interpreted, however, in a liberal sense, they imply, what we believe to be demonstrable, that the two had *a simultaneous origin*. The passage is either nonsense, or it is an admission that abstract and concrete mathematics are coeval. Thus, at the very first step, the alleged congruity between the order of generality and the order of evolution, does not hold good.

But may it not be that though abstract and concrete mathematics took their rise at the same time, the one afterwards developed, more rapidly than the other; and has ever since remained in advance of it? No: and again we call M. Comte himself as witness. Fortunately for his argument he has said nothing respecting the early stages of the concrete and abstract divisions after their divergence from a common root; otherwise the advent of Algebra long after the Greek geometry had reached a high development, would have been an inconvenient fact for him to deal with. But passing over this, and limiting ourselves to his own statements, we find, at the opening of the next chapter, the admission, that "the historical development of the abstract portion of mathematical science has, since the time of Descartes, been for the most part *determined* by that of the concrete." Further on we read respecting algebraic functions that "most functions were concrete in their origin — even those which are at present the most purely abstract; and the ancients discovered only through geometrical definitions elementary algebraic properties of functions to which a numerical value was not attached till long afterwards, rendering abstract to us what was concrete to the old geometers." How do these statements tally with his doctrine? Again, having divided the calculus into algebraic and arithmetical, M. Comte admits, as perforce he must, that the algebraic is more general than the arithmetical; yet he will not say that algebra preceded arithmetic in point of time. And again, having divided the calculus of functions into the calculus of direct functions (common algebra) and the calculus of indirect functions (transcendental analysis), he is obliged to speak of this last as possessing a higher generality than the first; yet it is far more modern. Indeed, by implication, M. Comte himself confesses this incongruity; for he says: — "It might seem that the transcendental analysis ought to be studied before the ordinary, as it provides the equations which the other has to resolve; but though the transcendental *is logically independent of the ordinary*, it is best to follow the usual method of study, taking the ordinary first." In all these cases, then, as well as at the close of the section where he predicts that mathematicians will in time "create procedures of *a wider generality*," M. Comte makes admissions that are diametrically opposed to the alleged law.

In the succeeding chapters treating of the concrete department of mathematics, we find similar contradictions. M. Comte himself names the geometry of the ancients *special* geometry, and that of moderns the *general* geometry. He admits that while "the ancients studied geometry with reference to the *bodies* under notice, or specially; the moderns study it with reference to the *phenomena* to be considered, or generally." He admits that while "the ancients extracted all they could out of one line or surface before passing to another," "the moderns, since Descartes, employ themselves on questions which relate to any figure whatever." These facts are the reverse of what, according to his theory, they should be. So, too, in mechanics. Before dividing it into statics and dynamics, M. Comte treats of the three laws of *motion,* and is obliged to do so; for statics, the more *general* of the two divisions, though it does not involve motion, is impossible as a science until the laws of motion are ascertained. Yet the laws of motion pertain to dynamics, the more *special* of the divisions. Further on he points out that after Archimedes, who discovered the law of equilibrium of the lever, statics made no progress until the establishment of dynamics enabled us to seek "the conditions of equilibrium through the laws of the composition of forces." And he adds — "At this day *this is the method universally employed.* At the first glance it does not appear the most rational — dynamics being more complicated than statics, and precedence being natural to the simpler. It would, in fact, be more philosophical to refer dynamics to statics, as has since been done." Sundry discoveries are afterwards detailed, showing how completely the development of statics has been achieved by considering its problems dynamically; and before the close of the section M. Comte remarks that "before hydrostatics could be comprehended under statics, it was necessary that the abstract theory of equilibrium, should be made so general as to apply directly to fluids as well as solids. This was accomplished when Lagrange supplied, as the basis of the whole of rational mechanics, the single principle of virtual velocities." In which statement we have two facts directly at variance with M. Comte's doctrine; — first, that the simpler science, statics, reached its present development only by the aid of the principle of virtual velocities, which belongs to the more complex science, dynamics; and that this "single principle" underlying all rational mechanics — this *most general form* which includes alike the relations of statical, hydrostatical, and dynamical forces — was reached so late as the time of Lagrange.

Thus it is *not* true that the historical succession of the divisions of mathematics has corresponded with the order of decreasing generality. It is *not* true that abstract mathematics was evolved antecedently to, and independently of concrete mathematics. It is *not* true that of the subdivisions of abstract mathematics, the more general came before the more special. And it is *not* true that concrete mathematics, in either of its two sections, began with the most abstract and advanced to the less abstract truths.

It may be well to mention, parenthetically, that in defending his alleged law of progression from the general to the special, M. Comte somewhere comments upon the two meanings of the word *general*, and the resulting liability to confusion. Without now discussing whether the asserted distinction can be maintained in other cases, it is manifest that it does not exist here. In sundry of the instances above quoted, the endeavors made by M. Comte himself to disguise, or to explain away, the precedence of the special over the general, clearly indicate that the generality spoken of, is of the kind meant by his formula. And it needs but a brief consideration of the matter to show that, even did he attempt it, he could not distinguish this generality, which, as above proved, frequently comes last, from the generality which he says always comes first. For what is the nature of that mental process by which objects, dimensions, weights, times, and the rest, are found capable of having their relations expressed numerically? It is the formation of certain abstract conceptions of unity, duality and multiplicity, which are applicable to all things alike. It is the invention of general symbols serving to express the numerical relations of entities, whatever be their special characters. And what is the nature of the mental process by which numbers are found capable of having their relations expressed algebraically? It is just the same. It is the formation of certain abstract conceptions of numerical functions which are the same whatever be the magnitudes of the numbers. It is the invention of general symbols serving to express the relations between numbers, as numbers express the relations between things. And transcendental analysis stands to algebra in the same position that algebra stands in to arithmetic.

To briefly illustrate their respective powers; — arithmetic can express in one formula the value of a *particular* tangent to a *particular* curve; algebra can express in one formula the values of *all* tangents to a *particular* curve; transcendental analysis can express in one formula the values of *all* tangents to *all* curves.

Just as arithmetic deals with the common properties of lines, areas, bulks, forces, periods; so does algebra deal with the common properties of the numbers which arithmetic presents; so does transcendental analysis deal with the common properties of the equations exhibited by algebra. Thus, the generality of the higher branches of the calculus, when compared with the lower, is the same kind of generality as that of the lower branches when compared with geometry or mechanics. And on examination it will be found that the like relation exists in the various other cases above given.

Having shown that M. Comte's alleged law of progression does not hold among the several parts of the same science, let us see how it agrees with the facts when applied to separate sciences. "Astronomy," says M. Comte, at the opening of Book III., "was a positive science, in its geometrical aspect, from the earliest days of the school of Alexandria; but Physics, which we are now to consider, had no positive character at all till Galileo made his great discoveries on the fall of heavy bodies." On this, our comment is simply that it is a misrepresentation based upon an arbitrary misuse of words — a mere verbal artifice. By choosing to exclude from terrestrial physics those laws of magnitude, motion, and position, which he includes in celestial physics, M. Comte makes it appear that the one owes nothing to the other. Not only is this altogether unwarrantable, but it is radically inconsistent with his own scheme of divisions. At the outset he says — and as the point is important we quote from the original — "Pour la *physique inorganique* nous voyons d'abord, en nous conformant tonjours à l'ordre de généralité et de dépendance des phénomènes, qu'elle doit être partagée en deux sections distinctes, suivant qu'elle considère les phénomènes généraux de l'univers, ou, en particulier, ceux que présentent les corps terrestres. D'ou la physique céleste, ou l'astronomie, soit géométrique, soit mechanique; et la physique terrestre."

Here then we have *inorganic physics* clearly divided into *celestial physics* and *terrestrial physics* — the phenomena presented by the universe, and the phenomena presented by earthly bodies. If now celestial bodies and terrestrial bodies exhibit sundry leading phenomena in common, as they do, how can the generalization of these common phenomena be considered as pertaining to the one class rather than to the other? If inorganic physics includes geometry (which M. Comte has made it do by comprehending *geometrical* astronomy in its sub- section — celestial physics); and if its sub-section — terrestrial physics, treats of things having geometrical properties; how can the laws of geometrical relations be excluded from terrestrial physics? Clearly if celestial physics includes the geometry of objects in the heavens, terrestrial physics includes the geometry of objects on the earth. And if terrestrial physics includes terrestrial geometry, while celestial physics includes celestial geometry, then the geometrical part of terrestrial physics precedes the geometrical part of celestial physics; seeing that geometry gained its first ideas from surrounding objects. Until men had learnt geometrical relations from bodies on the earth, it was impossible for them to understand the geometrical relations of bodies in the heavens.

So, too, with celestial mechanics, which had terrestrial mechanics for its parent. The very conception of *force*, which underlies the whole of mechanical astronomy, is borrowed from our earthly experiences, and the leading laws of mechanical action as exhibited in scales, levers, projectiles, &c., had to be ascertained before the dynamics of the solar system could be entered upon. What were the laws made use of by Newton in working out his grand discovery? The law of falling bodies disclosed by Galileo; that of the composition of forces also disclosed by Galileo; and that of centrifugal force found out by Huyghens — all of them generalizations of terrestrial physics. Yet, with facts like these before him, M. Comte places astronomy before physics in order of evolution! He does not compare the geometrical parts of the two together, and the mechanical parts of the two together; for this would by no means suit his hypothesis. But he compares the geometrical part of the one with the mechanical part of the other, and so gives a semblance of truth to his position. He is led away by a verbal delusion. Had he confined his attention to the things and disregarded the words, he would have seen that before mankind scientifically co-ordinated *any one class of phenomena* displayed in the heavens, they had previously co-ordinated a *parallel class of phenomena* displayed upon the surface of the earth.

Were it needful we could fill a score pages with the incongruities of M. Comte's scheme. But the foregoing samples will suffice. So far is his law of evolution of the sciences from being tenable, that, by following his example, and arbitrarily ignoring one class of facts, it would be possible to present, with great plausibility, just the opposite generalization to that which he enunciates. While he asserts that the rational order of the sciences, like the order of their historic development, "is determined by the degree of simplicity, or, what comes to the same thing, of generality of their phenomena;" it might contrariwise be asserted, that, commencing with the complex and the special, mankind have progressed step by step to a knowledge of greater simplicity and wider generality. So much evidence is there of this as to have drawn from Whewell, in his *History of the Inductive Sciences*, the general remark that "the reader has already seen repeatedly in the course of this history, complex and derivative principles presenting themselves to men's minds before simple and elementary ones."

Even from M. Comte's own work, numerous facts, admissions, and arguments, might be picked out, tending to show this. We have already quoted his words in proof that both abstract and concrete mathematics have progressed towards a higher degree of generality, and that he looks forward to a higher generality still. Just to strengthen this adverse hypothesis, let us take a further instance. From the *particular* case of the scales, the law of equilibrium of which was familiar to the earliest nations known, Archimedes advanced to the more *general* case of the unequal lever with unequal weights; the law of equilibrium of which *includes* that of the scales. By the help of Galileo's discovery concerning the composition of forces, D'Alembert "established, for the first time, the equations of equilibrium of *any* system of forces applied to the different points of a solid body" — equations which include all cases of levers and an infinity of cases besides. Clearly this is progress towards a higher generality — towards a knowledge more independent of special circumstances — towards a study of phenomena "the most disengaged from the incidents of particular cases;" which is M. Comte's definition of "the most simple phenomena." Does it not indeed follow from the familiarly admitted fact, that mental advance is from the concrete to the abstract, from the particular to the general, that the universal and therefore most simple truths are the last to be discovered? Is not the government of the solar system by a force varying inversely as the square of the distance, a simpler conception than any that preceded it? Should we ever succeed in reducing all orders of phenomena to some single law — say of atomic action, as M. Comte suggests — must not that law answer to his test of being *independent* of all others, and therefore most simple? And would not such a law generalize the phenomena of gravity, cohesion, atomic affinity, and electric repulsion, just as the laws of number generalize the quantitative phenomena of space, time and force?

The possibility of saying so much in support of an hypothesis the very reverse of M. Comte's, at once proves that his generalization is only a half-truth. The fact is, that neither proposition is correct by itself; and the actuality is expressed only by putting the two together. The progress of science is duplex: it is at once from the special to the general, and from the general to the special: it is analytical and synthetical at the same time.

M. Comte himself observes that the evolution of science has been accomplished by the division of labour; but he quite misstates the mode in which this division of labour has operated. As he describes it, it has simply been an arrangement of phenomena into classes, and the study of each class by itself. He does not recognise the constant effect of progress in each class upon *all* other classes; but only on the class succeeding it in his hierarchical scale. Or if he occasionally admits collateral influences and intercommunications, he does it so grudgingly, and so quickly puts the admissions out of sight and forgets them, as to leave the impression that, with but trifling exceptions, the sciences aid each other only in the order of their alleged succession. The fact is, however, that the division of labour in science, like the division of labour in society, and like the "physiological division of labour" in individual organisms, has been not only a specialization of functions, but a continuous helping of each division by all the others, and of all by each. Every particular class of inquirers has, as it were, secreted its own particular order of truths from the general mass of material which observation accumulates; and all other classes of inquirers have made use of these truths as fast as they were elaborated, with the effect of enabling them the better to elaborate each its own order of truths.

It was thus in sundry of the cases we have quoted as at variance with M. Comte's doctrine. It was thus with the application of Huyghens's optical discovery to astronomical observation by

Galileo. It was thus with the application of the isochronism of the pendulum to the making of instruments for measuring intervals, astronomical and other. It was thus when the discovery that the refraction and dispersion of light did not follow the same law of variation, affected both astronomy and physiology by giving us achromatic telescopes and microscopes. It was thus when Bradley's discovery of the aberration of light enabled him to make the first step towards ascertaining the motions of the stars. It was thus when Cavendish's torsion-balance experiment determined the specific gravity of the earth, and so gave a datum for calculating the specific gravities of the sun and planets. It was thus when tables of atmospheric refraction enabled observers to write down the real places of the heavenly bodies instead of their apparent places. It was thus when the discovery of the different expansibilities of metals by heat, gave us the means of correcting our chronometrical measurements of astronomical periods. It was thus when the lines of the prismatic spectrum were used to distinguish the heavenly bodies that are of like nature with the sun from those which are not. It was thus when, as recently, an electro-telegraphic instrument was inverted for the more accurate registration of meridional transits. It was thus when the difference in the rates of a clock at the equator, and nearer the poles, gave data for calculating the oblateness of the earth, and accounting for the precession of the equinoxes. It was thus — but it is needless to continue.

Here, within our own limited knowledge of its history, we have named ten additional cases in which the single science of astronomy has owed its advance to sciences coming *after* it in M. Comte's series. Not only its secondary steps, but its greatest revolutions have been thus determined. Kepler could not have discovered his celebrated laws had it not been for Tycho Brahe's accurate observations; and it was only after some progress in physical and chemical science that the improved instruments with which those observations were made, became possible. The heliocentric theory of the solar system had to wait until the invention of the telescope before it could be finally established. Nay, even the grand discovery of all — the law of gravitation — depended for its proof upon an operation of physical science, the measurement of a degree on the Earth's surface. So completely indeed did it thus depend, that Newton *had actually abandoned his hypothesis* because the length of a degree, as then stated, brought out wrong results; and it was only after Picart's more exact measurement was published, that he returned to his calculations and proved his great generalization. Now this constant intercommunion, which, for brevity's sake, we have illustrated in the case of one science only, has been taking place with all the sciences. Throughout the whole course of their evolution there has been a continuous *consensus* of the sciences — a *consensus* exhibiting a general correspondence with the *consensus* of faculties in each phase of mental development; the one being an objective registry of the subjective state of the other.

From our present point of view, then, it becomes obvious that the conception of a *serial* arrangement of the sciences is a vicious one. It is not simply that the schemes we have examined are untenable; but it is that the sciences cannot be rightly placed in any linear order whatever. It is not simply that, as M. Comte admits, a classification "will always involve something, if not arbitrary, at least artificial;" it is not, as he would have us believe, that, neglecting minor imperfections a classification may be substantially true; but it is that any grouping of the sciences in a succession gives a radically erroneous idea of their genesis and their dependencies. There is no "one *rational* order among a host of possible systems." There is no "true *filiation* of the sciences." The whole hypothesis is fundamentally false. Indeed, it needs but a glance at its origin to see at once how baseless it is. Why a *series?* What reason have we to suppose that the sciences admit of a *linear* arrangement? Where is our warrant for assuming that there is some *succession* in which they can be placed? There is no reason; no warrant. Whence then has arisen the supposition? To use M. Comte's own phraseology, we should say, it is a metaphysical conception. It adds another to the cases constantly occurring, of the human mind being made the measure of Nature. We are obliged to think in sequence; it is the law of our minds that we must consider subjects separately, one after another: *therefore* Nature must be serial — *therefore* the sciences must be classifiable in a succession. See here the birth of the notion, and the sole evidence of its truth. Men have been obliged when arranging in books their schemes of education and systems of knowledge, to choose *some* order or other. And from inquiring what is the best order, have naturally fallen into the belief that there is an order which truly represents the facts — have persevered in seeking such an order; quite overlooking the previous question whether it is likely that Nature has consulted the convenience of book-making.

For German philosophers, who hold that Nature is "petrified intelligence," and that logical forms are the foundations of all things, it is a consistent hypothesis that as thought is serial, Nature is serial; but that M. Comte, who is so bitter an opponent of all anthropomorphism, even in its most evanescent shapes, should have committed the mistake of imposing upon the external world an arrangement which so obviously springs from a limitation of the human consciousness, is somewhat strange. And it is the more strange when we call to mind how, at the outset, M. Comte remarks that in the beginning "*toutes les sciences sont cultivées simultanément par les mêmes esprits;*" that this is "*inevitable et même indispensable;*" and how he further remarks that the different sciences are "*comme les diverses branches d'un tronc unique.*" Were it not accounted for by the distorting influence of a cherished hypothesis, it would be scarcely possible to understand how, after recognising truths like these, M. Comte should have persisted in attempting to construct "*une échelle encyclopédique.*"

The metaphor which M. Comte has here so inconsistently used to express the relations of the sciences — branches of one trunk — is an approximation to the truth, though not the truth itself. It suggests the facts that the sciences had a common origin; that they have been developing simultaneously; and that they have been from time to time dividing and sub-dividing. But it does not suggest the yet more important fact, that the divisions and sub-divisions thus arising do not remain separate, but now and again re-unite in direct and indirect ways. They mosculate; they severally send off and receive connecting growths; and the intercommunion has been ever becoming more frequent, more intricate, more widely ramified. There has all along been higher specialization, that. there might be a larger generalization; and a deeper analysis, that there might be a better synthesis. Each larger generalization has lifted sundry specializations still higher; and each better synthesis has prepared the way for still deeper analysis.

And here we may fitly enter upon the task awhile since indicated — a sketch of the Genesis of Science, regarded as a gradual outgrowth from common knowledge — an extension of the perceptions by the aid of the reason. We propose to treat it as a psychological process historically displayed; tracing at the same time the advance from qualitative to quantitative prevision; the progress from concrete facts to abstract facts, and the application of such abstract facts to the analysis of new orders of concrete facts; the simultaneous advance in gereralization and specialization; the continually increasing subdivision and reunion of the sciences; and their constantly improving *consensus*.

To trace out scientific evolution from its deepest roots would, of course, involve a complete analysis of the mind. For as science is a development of that common knowledge acquired by the unaided senses and uncultured reason, so is that common knowledge itself gradually built up out of the simplest perceptions. We must, therefore, begin somewhere abruptly; and the most appropriate stage to take for our point of departure will be the adult mind of the savage.

Commencing thus, without a proper preliminary analysis we are naturally somewhat at a loss how to present, in a satisfactory manner, those fundamental processes of thought out of which science ultimately originates. Perhaps our argument may be best initiated by the proposition, that all intelligent action whatever depends upon the discerning of distinctions among surrounding things. The condition under which only it is possible for any creature to obtain food and avoid danger is, that it shall be differently affected by different objects — that it shall be led to act in one way by one object, and in another way by another. In the lower orders of creatures this condition is fulfilled by means of an apparatus which acts automatically. In the higher orders the actions are partly automatic, partly conscious. And in man they are almost wholly conscious.

Throughout, however, there must necessarily exist a certain classification of things according to their properties — a classification which is either organically registered in the system, as in the inferior creation, or is formed by experience, as in ourselves. And it may be further remarked, that the extent to which this classification is carried, roughly indicates the height of intelligence — that, while the lowest organisms are able to do little more than discriminate organic from inorganic matter; while the generality of animals carry their classifications no further than to a limited number of plants or creatures serving for food, a limited number of beasts of prey, and a limited number of places and materials; the most degraded of the human race possess a knowledge of the distinctive natures of a great variety of substances, plants, animals, tools, persons, &c., not only as classes but as individuals.

What now is the mental process by which classification is effected? Manifestly it is a recognition of the *likeness* or *unlikeness* of things, either in respect of their sizes, colours, forms, weights, textures, tastes, &c., or in respect of their modes of action. By some special mark, sound, or motion, the savage identifies a certain four-legged creature he sees, as one that is good for food, and to be caught in a particular way; or as one that is dangerous; and acts accordingly. He has classed together all the creatures that are *alike* in this particular. And manifestly in choosing the wood out of which to form his bow, the plant with which to poison his arrows, the bone from which to make his fish-hooks, he identifies them through their chief sensible properties as belonging to the general classes, wood, plant, and bone, but distinguishes them as belonging to sub-classes by virtue of certain properties in which they are *unlike* the rest of the general classes they belong to; and so forms genera and species.

And here it becomes manifest that not only is classification carried on by grouping together in the mind things that are *like;* but that classes and sub-classes are formed and arranged according to the *degrees of unlikeness.* Things widely contrasted are alone distinguished in the lower stages of mental evolution; as may be any day observed in an infant. And gradually as the powers of discrimination increase, the widely contrasted classes at first distinguished, come to be each divided into sub-classes, differing from each other less than the classes differ; and these sub-classes are again divided after the same manner. By the continuance of which process, things are gradually arranged into groups, the members of which are less and less *unlike;* ending, finally, in groups whose members differ only as individuals, and not specifically. And thus there tends ultimately to arise the notion of *complete likeness.* For manifestly, it is impossible that groups should continue to be sub-divided in virtue of smaller and smaller differences, without there being a simultaneous approximation to the notion of *no difference.*

Let us next notice that the recognition of likeness and unlikeness, which underlies classification, and out of which continued classification evolves the idea of complete likeness — let us next notice that it also underlies the process of *naming,* and by consequence *language.* For all language consists, at the beginning, of symbols which are as *like* to the things symbolized as it is practicable to make them. The language of signs is a means of conveying ideas by mimicking the actions or peculiarities of the things referred to. Verbal language is also, at the beginning, a mode of suggesting objects or acts by imitating the sounds which the objects make, or with which the acts are accompanied. Originally these two languages were used simultaneously. It needs but to watch the gesticulations with which the savage accompanies his speech — to see a Bushman or a Kaffir dramatizing before an audience his mode of catching game — or to note the extreme paucity of words in all primitive vocabularies; to infer that at first, attitudes, gestures, and sounds, were all combined to produce as good a *likeness* as possible, of the things, animals, persons, or events described; and that as the sounds came to be understood by themselves the gestures fell into disuse: leaving traces, however, in the manners of the more excitable civilized races. But be this as it may, it suffices simply to observe, how many of the words current among barbarous peoples are like the sounds appertaining to the things signified; how many of our own oldest and simplest words have the same peculiarity; how children tend to invent imitative words; and how the sign-language spontaneously formed by deaf mutes is invariably based upon imitative actions — to at once see that the notion of *likeness* is that from which the nomenclature of objects takes its rise.

Were there space we might go on to point out how this law of life is traceable, not only in the origin but in the development of language; how in primitive tongues the plural is made by a duplication of the singular, which is a multiplication of the word to make it *like* the multiplicity of the things; how the use of metaphor — that prolific source of new words — is a suggesting of ideas that are *like* the ideas to be conveyed in some respect or other; and how, in the copious use of simile, fable, and allegory among uncivilized races, we see that complex conceptions, which there is yet no direct language for, are rendered, by presenting known conceptions more or less *like* them.

This view is further confirmed, and the predominance of this notion of likeness in primitive times further illustrated, by the fact that our system of presenting ideas to the eye originated after the same fashion. Writing and printing have descended from picture-language. The earliest mode of permanently registering a fact was by depicting it on a wall; that is — by exhibiting something as *like* to the thing to be remembered as it could be made. Gradually as the practice grew habitual and extensive, the most frequently repeated forms became fixed, and presently abbreviated; and, passing through the hieroglyphic and ideographic phases, the symbols lost all apparent relations to the things signified: just as the majority of our spoken words have done.

Observe again, that the same thing is true respecting the genesis of reasoning. The *likeness* that is perceived to exist between cases, is the essence of all early reasoning and of much of our present reasoning. The savage, having by experience discovered a relation between a certain object and a certain act, infers that the *like* relation will be found in future cases. And the expressions we constantly use in our arguments — "*analogy* implies," "the cases are *not parallel,*" "*by parity of* reasoning," "there is no *similarity,*" — show how constantly the idea of likeness underlies our ratiocinative processes.

Still more clearly will this be seen on recognising the fact that there is a certain parallelism between reasoning and classification; that the two have a common root; and that neither can go on without the other. For on the one hand, it is a familiar truth that the attributing to a body hi consequence of some of its properties, all those other properties in virtue of which it is referred to a particular class, is an act of inference. And, on the other hand, the forming of a generalization is the putting together in one class, all those cases which present like relations; while the drawing a deduction is essentially the perception that a particular case belongs to a certain class of cases previously generalized. So that as classification is a grouping together of *like things*; reasoning is a grouping together of *like relations* among things. Add to which, that while the perfection gradually achieved in classification consists in the formation of groups of *objects* which are *completely alike;* the perfection gradually achieved in reasoning consists in the formation of groups of *cases* which are *completely alike.*

Once more we may contemplate this dominant idea of likeness as exhibited in art. All art, civilized as well as savage, consists almost wholly in the making of objects *like* other objects; either as found in Nature, or as produced by previous art. If we trace back the varied art-products now existing, we find that at each stage the divergence from previous patterns is but small when compared with the agreement; and in the earliest art the persistency of imitation is yet more conspicuous. The old forms and ornaments and symbols were held sacred, and perpetually copied. Indeed, the strong imitative tendency notoriously displayed by the lowest human races, ensures among them a constant reproducing of likenesses of things, forms, signs, sounds, actions, and whatever else is imitable; and we may even suspect that this aboriginal peculiarity is in some way connected with the culture and development of this general conception, which we have found so deep and widespread in its applications.

And now let us go on to consider how, by a further unfolding of this same fundamental notion, there is a gradual formation of the first germs of science. This idea of likeness which underlies classification, nomenclature, language spoken and written, reasoning, and art; and which plays so important a part because all acts of intelligence are made possible only by distinguishing among surrounding things, or grouping them into like and unlike; — this idea we shall find to be the one of which science is the especial product. Already during the stage we have been describing, there has existed *qualitative* prevision in respect to the commoner phenomena with which savage life is familiar; and we have now to inquire how the elements of *quantitative* prevision are evolved. We shall find that they originate by the perfecting of this same idea of likeness; that they have their rise in that conception of *complete likeness* which, as we have seen, necessarily results from the continued process of classification.

For when the process of classification has been carried as far as it is possible for the uncivilized to carry it — when the animal kingdom has been grouped not merely into quadrupeds, birds, fishes, and insects, but each of these divided into lands — when there come to be sub-classes, in each of which the members differ only as individuals, and not specifically; it is clear that there must occur a frequent observation of objects which differ so little as to be indistinguishable. Among several creatures which the savage has killed and carried home, it must often happen that some one, which he wished to identify, is so exactly like another that he cannot tell which is which. Thus, then, there originates the notion of *equality*. The things which among ourselves are called *equal* — whether lines, angles, weights, temperatures, sounds or colours — are things which produce in us sensations that cannot be distinguished from each other. It is true that we now apply the word *equal* chiefly to the separate phenomena which objects exhibit, and not to groups of phenomena; but this limitation of the idea has evidently arisen by subsequent analysis. And that the notion of equality did thus originate, will, we think, become obvious on remembering that as there were no artificial objects from which it could have been abstracted, it must have been abstracted from natural objects; and that the various families of the animal kingdom chiefly furnish those natural objects which display the requisite exactitude of likeness.

The same order of experiences out of which this general idea of equality is evolved, gives birth at the same time to a more complex idea of equality; or, rather, the process just described generates an idea of equality which further experience separates into two ideas — *equality of things* and *equality of relations*. While organic, and more especially animal forms, occasionally exhibit this perfection of likeness out of which the notion of simple equality arises, they more frequently exhibit only that kind of likeness which we call *similarity;* and which is really compound equality. For the similarity of two creatures of the same species but of different sizes, is of the same nature as the similarity of two geometrical figures. In either case, any two parts of the one bear the same ratio to one another, as the homologous parts of the other. Given in any species, the proportions found to exist among the bones, and we may, and zoologists do, predict from any one, the dimensions of the rest; just as, when knowing the proportions subsisting among the parts of a geometrical figure, we may, from the length of one, calculate the others. And if, in the case of similar geometrical figures, the similarity can be established only by proving exactness of proportion among the homologous parts; if we express this relation between two parts in the one, and the corresponding parts in the other, by the formula A is to B as *a* is to *b*; if we otherwise write this, A to B = a to *b;* if, consequently, the fact we prove is that the relation of A to B *equals* the relation of a to *b*; then it is manifest that the fundamental conception of similarity is *equality of relations.*

With this explanation we shall be understood when we say that the notion of equality of relations is the basis of all exact reasoning. Already it has been shown that reasoning in general is a recognition of *likeness* of relations; and here we further find that while the notion of likeness of things ultimately evolves the idea of simple equality, the notion of likeness of relations evolves the idea of equality of relations: of which the one is the concrete germ of exact science, while the other is its abstract germ.

Those who cannot understand how the recognition of similarity in creatures of the same kind, can have any alliance with reasoning, will get over the difficulty on remembering that the phenomena among which equality of relations is thus perceived, are phenomena of the same order and are present to the senses at the same time; while those among which developed reason perceives relations, are generally neither of the same order, nor simultaneously present. And if further, they will call to mind how Cuvier and Owen, from a single part of a creature, as a tooth, construct the rest by a process of reasoning based on this equality of relations, they will see that the two things are intimately connected, remote as they at first seem. But we anticipate. What it concerns us here to observe is, that from familiarity with organic forms there simultaneously arose the ideas of *simple equality,* and *equality of relations.*

At the same time, too, and out of the same mental processes, came the first distinct ideas of *number.* In the earliest stages, the presentation of several like objects produced merely an indefinite conception of multiplicity; as it still does among Australians, and Bushmen, and Damaras, when the number presented exceeds three or four. With such a fact before us we may safely infer that the first clear numerical conception was that of duality as contrasted with unity. And this notion of duality must necessarily have grown up side by side with those of likeness and equality; seeing that it is impossible to recognise the likeness of two things without also perceiving that there are two. From the very beginning the conception of number must have been, as it is still, associated with the likeness or equality of the things numbered. If we analyze it, we find that simple enumeration is a registration of repeated impressions of any kind. That these may be capable of enumeration it is needful that they be more or less alike; and before any *absolutely true* numerical results can be reached, it is requisite that the units be *absolutely equal.* The only way in which we can establish a numerical relationship between things that do not yield us like impressions, is to divide them into parts that *do* yield us like impressions. Two unlike magnitudes of extension, force, time, weight, or what not, can have their relative amounts estimated, only by means of some small unit that is contained many times in both; and even if we finally write down the greater one as a unit and the other as a fraction of it, we state, in the denominator of the fraction, the number of parts into which the unit must be divided to be comparable with the fraction.

It is, indeed, true, that by an evidently modern process of abstraction, we occasionally apply numbers to unequal units, as the furniture at a sale or the various animals on a farm, simply as so many separate entities; but no true result can be brought out by calculation with units of this order. And, indeed, it is the distinctive peculiarity of the calculus in general, that it proceeds on the hypothesis of that absolute equality of its abstract units, which no real units possess; and that the exactness of its results holds only in virtue of this hypothesis. The first ideas of number must necessarily then have been derived from like or equal magnitudes as seen chiefly in organic objects; and as the like magnitudes most frequently observed were magnitudes of extension, it follows that geometry and arithmetic had a simultaneous origin.

Not only are the first distinct ideas of number co-ordinate with ideas of likeness and equality, but the first efforts at numeration displayed the same relationship. On reading the accounts of various savage tribes, we find that the method of counting by the fingers, still followed by many children, is the aboriginal method. Neglecting the several cases in which the ability to enumerate does not reach even to the number of fingers on one hand, there are many cases in which it does not extend beyond ten — the limit of the simple finger notation. The fact that in so many instances, remote, and seemingly unrelated nations, have adopted *ten* as their basic number; together with the fact that in the remaining instances the basic number is either *five* (the fingers of one hand) or *twenty* (the fingers and toes); almost of themselves show that the fingers were the original units of numeration. The still surviving use of the word *digit,* as the general name for a figure in arithmetic, is significant; and it is even said that our word *ten* (Sax. tyn; Dutch, tien; German, zehn) means in its primitive expanded form *two hands.* So that originally, to say there were ten things, was to say there were two hands of them.

From all which evidence it is tolerably clear that the earliest mode of conveying the idea of any number of things, was by holding up as many fingers as there were things; that is — using a symbol which was *equal,* in respect of multiplicity, to the group symbolized. For which inference there is, indeed, strong confirmation in the recent statement that our own soldiers are even now spontaneously adopting this device in their dealings with the Turks. And here it should be remarked that in this recombination of the notion of equality with that of multiplicity, by which the first steps in numeration are effected, we may see one of the earliest of those inosculations between the diverging branches of science, which are afterwards of perpetual occurrence.

Indeed, as this observation suggests, it will be well, before tracing the mode in which exact science finally emerges from the merely approximate judgments of the senses, and showing the non-serial evolution of its divisions, to note the non-serial character of those preliminary processes of which all after development is a continuation. On re-considering them it will be seen that not only are they divergent growths from a common root, — not only are they simultaneous in their progress; but that they are mutual aids; and that none can advance without the rest. That completeness of classification for which the unfolding of the perceptions paves the way, is impossible without a corresponding progress in language, by which greater varieties of objects are thinkable and expressible. On the one hand it is impossible to carry classification far without names by which to designate the classes; and on the other hand it is impossible to make language faster than things are classified.

Again, the multiplication of classes and the consequent narrowing of each class, itself involves a greater likeness among the things classed together; and the consequent approach towards the notion of complete likeness itself allows classification to be carried higher. Moreover, classification necessarily advances *pari passu* with rationality — the classification of *things* with the classification of *relations.* For things that belong to the same class are, by implication, things of which the properties and modes of behaviour — the co-existences and sequences — are more or less the same; and the recognition of this sameness of co-existences and sequences is reasoning. Whence it follows that the advance of classification is necessarily proportionate to the advance of generalizations. Yet further, the notion of *likeness,* both in things and relations, simultaneously evolves by one process of culture the ideas of *equality* of things and *equality* of relations; which are the respective bases of exact concrete reasoning and exact abstract reasoning — Mathematics and Logic. And once more, this idea of equality, in the very process of being formed, necessarily gives origin to two series of relations — those of magnitude and those of number: from which arise geometry and the calculus. Thus the process throughout is one of perpetual subdivision and perpetual intercommunication of the divisions. From the very first there has been that *consensus* of different kinds of knowledge, answering to the *consensus* of the intellectual faculties, which, as already said, must exist among the sciences.

Let us now go on to observe how, out of the notions of *equality* and *number,* as arrived at in the manner described, there gradually arose the elements of quantitative prevision.

Equality, once having come to be definitely conceived, was readily applicable to other phenomena than those of magnitude. Being predicable of all things producing indistinguishable impressions, there naturally grew up ideas of equality in weights, sounds, colours, &c.; and indeed it can scarcely be doubted that the occasional experience of equal weights, sounds, and colours, had a share in developing the abstract conception of equality — that the ideas of equality in size, relations, forces, resistances, and sensible properties in general, were evolved during the same period. But however this may be, it is clear that as fast as the notion of equality gained definiteness, so fast did that lowest kind of quantitative prevision which is achieved without any instrumental aid, become possible.

The ability to estimate, however roughly, the amount of a foreseen result, implies the conception that it will be *equal to* a certain imagined quantity; and the correctness of the estimate will manifestly depend upon the accuracy at which the perceptions of sensible equality have arrived. A savage with a piece of stone in his hand, and another piece lying before him of greater bulk but of the same kind (a fact which he infers from the *equality* of the two in colour and texture) knows about what effort he must put forth to raise this other piece; and he judges accurately in proportion to the accuracy with which he perceives that the one is twice, three times, four times, &c. as large as the other; that is — in proportion to the precision of his ideas of equality and number. And here let us not omit to notice that even in these vaguest of quantitative previsions, the conception of *equality of relations* is also involved. For it is only in virtue of an undefined perception that the relation between bulk and weight in the one stone is *equal to* the relation between bulk and weight in the other, that even the roughest approximation can be made.

But how came the transition from those uncertain perceptions of equality which the unaided senses give, to the certain ones with which science deals? It came by placing the things compared in juxtaposition. Equality being predicated of things which give us indistinguishable impressions, and no accurate comparison of impressions being possible unless they occur in immediate succession, it results that exactness of equality is ascertainable in proportion to the closeness of the compared things. Hence the fact that when we wish to judge of two shades of colour whether they are alike or not, we place them side by side; hence the fact that we cannot, with any precision, say which of two allied sounds is the louder, or the higher in pitch, unless we hear the one immediately after the other; hence the fact that to estimate the ratio of weights, we take one in each hand, that we may compare their pressures by rapidly alternating in thought from the one to the other; hence the fact, that in a piece of music, we can continue to make equal beats when the first beat has been given, but cannot ensure commencing with the same length of beat on a future occasion; and hence, lastly, the fact, that of all magnitudes, those of *linear extension* are those of which the equality is most accurately ascertainable, and those to which by consequence all others have to be reduced. For it is the peculiarity of linear extension that it alone allows its magnitudes to be placed in *absolute* juxtaposition, or, rather, in coincident position; it alone can test the equality of two magnitudes by observing whether they will coalesce, as two equal mathematical lines do, when placed between the same points; it alone can test *equality* by trying whether it will become *identity*. Hence, then, the fact, that all exact science is reducible, by an ultimate analysis, to results measured in equal units of linear extension.

Still it remains to be noticed in what manner this determination of equality by comparison of linear magnitudes originated. Once more may we perceive that surrounding natural objects supplied the needful lessons. From the beginning there must have been a constant experience of like things placed side by side — men standing and walking together; animals from the same herd; fish from the same shoal. And the ceaseless repetition of these experiences could not fail to suggest the observation, that the nearer together any objects were, the more visible became any inequality between them. Hence the obvious device of putting in apposition, things of which it was desired to ascertain the relative magnitudes. Hence the idea of *measure*. And here we suddenly come upon a group of facts which afford a solid basis to the remainder of our argument; while they also furnish strong evidence in support of the foregoing speculations. Those who look sceptically on this attempted rehabilitation of the earliest epochs of mental development, and who more especially think that the derivation of so many primary notions from organic forms is somewhat strained, will perhaps see more probability in the several hypotheses that have been ventured, on discovering that all measures of *extension* and *force* originated from the lengths and weights of organic bodies; and all measures of *time* from the periodic phenomena of either organic or inorganic bodies.

Thus, among linear measures, the cubit of the Hebrews was the *length of the forearm* from the elbow to the end of the middle finger; and the smaller scriptural dimensions are expressed in *hand-breadths* and *spans*. The Egyptian cubit, which was similarly derived, was divided into digits, which were *finger-*

breadths; and each finger-breadth was more definitely expressed as being equal to four *grains of barley* placed breadthwise. Other ancient measures were the orgyia or *stretch of the arms,* the *pace,* and the *palm.* So persistent has been the use of these natural units of length in the East, that even now some of the Arabs mete out cloth by the forearm. So, too, is it with European measures. The *foot* prevails as a dimension throughout Europe, and has done since the time of the Romans, by whom, also, it was used: its lengths in different places varying not much more than men's feet vary. The heights of horses are still expressed in *hands.* The inch is the length of the terminal joint of *the thumb*; as is clearly shown in France, where *pouce* means both thumb and inch. Then we have the inch divided into three *barley-corns.*

So completely, indeed, have these organic dimensions served as the substrata of all mensuration, that it is only by means of them that we can form any estimate of some of the ancient distances. For example, the length of a degree on the Earth's surface, as determined by the Arabian astronomers shortly after the death of Haroun-al-Raschid, was fifty-six of their miles. We know nothing of their mile further than that it was 4000 cubits; and whether these were sacred cubits or common cubits, would remain doubtful, but that the length of the cubit is given as twenty-seven inches, and each inch defined as the thickness of six barley-grains. Thus one of the earliest measurements of a degree comes down to us in barley-grains. Not only did organic lengths furnish those approximate measures which satisfied men's needs in ruder ages, but they furnished also the standard measures required in later times. One instance occurs in our own history. To remedy the irregularities then prevailing, Henry I. commanded that the ulna, or ancient ell, which answers to the modern yard, should be made of the exact length of *his own arm.*

Measures of weight again had a like derivation. Seeds seem commonly to have supplied the unit. The original of the carat used for weighing in India is *a small bean.* Our own systems, both troy and avoirdupois, are derived primarily from wheat-corns. Our smallest weight, the grain, is *a gram of wheat.* This is not a speculation; it is an historically registered fact. Henry III. enacted that an ounce should be the weight of 640 dry grains of wheat from the middle of the ear. And as all the other weights are multiples or sub-multiples of this, it follows that the grain of wheat is the basis of our scale. So natural is it to use organic bodies as weights, before artificial weights have been established, or where they are not to be had, that in some of the remoter parts of Ireland the people are said to be in the habit, even now, of putting a man into the scales, to serve as a measure for heavy commodities.

Similarly with time. Astronomical periodicity, and the periodicity of animal and vegetable life, are simultaneously used in the first stages of progress for estimating epochs. The simplest unit of time, the day, nature supplies ready made. The next simplest period, the mooneth or month, is also thrust upon men's notice by the conspicuous changes constituting a lunation. For larger divisions than these, the phenomena of the seasons, and the chief events from time to time occurring, have been used by early and uncivilized races. Among the Egyptians the rising of the Nile served as a mark. The New Zealanders were found to begin their year from the reappearance of the Pleiades above the sea. One of the uses ascribed to birds, by the Greeks, was to indicate the seasons by their migrations. Barrow describes the aboriginal Hottentot as denoting periods by the number of moons before or after the ripening of one of his chief articles of food. He further states that the Kaffir chronology is kept by the moon, and is registered by notches on sticks — the death of a favourite chief, or the gaining of a victory, serving for a new era. By which last fact, we are at once reminded that in early history, events are commonly recorded as occurring in certain reigns, and in certain years of certain reigns: a proceeding which practically made a king's reign a measure of duration.

And, as further illustrating the tendency to divide time by natural phenomena and natural events, it may be noticed that even by our own peasantry the definite divisions of months and years are but little used; and that they habitually refer to occurrences as "before sheep-shearing," or "after harvest," or "about the time when the squire died." It is manifest, therefore, that the more or less equal periods perceived in Nature gave the first units of measure for time; as did Nature's more or less equal lengths and weights give the first units of measure for space and force.

It remains only to observe, as further illustrating the evolution of quantitative ideas after this manner, that measures of value were similarly derived. Barter, in one form or other, is found among all but the very lowest human races. It is obviously based upon the notion of *equality of worth.* And as it gradually merges into trade by the introduction of some kind of currency, we find that the *measures of worth,* constituting this currency, are organic bodies; in some cases *cowries,* in others *cocoa-nuts,* in others *cattle,* in others *pigs;* among the American Indians peltry or *skins,* and in Iceland *dried fish.*

Notions of exact equality and of measure having been reached, there came to be definite ideas of relative magnitudes as being multiples one of another; whence the practice of measurement by direct apposition of a measure. The determination of linear extensions by this process can scarcely be called science, though it is a step towards it; but the determination of lengths of time by an analogous process may be considered as one of the earliest samples of quantitative prevision. For when it is first ascertained that the moon completes the cycle of her changes in about thirty days — a fact known to most uncivilized tribes that can count beyond the number of their fingers — it is manifest that it becomes possible to say in what number of days any specified phase of the moon will recur; and it is also manifest that this prevision is effected by an opposition of two times, after the same manner that linear space is measured by the opposition of two lines. For to express the moon's period in days, is to say how many of these units of measure are contained in the period to be measured — is to ascertain the distance between two points in time by means of a *scale of days,* just as we ascertain the distance between two points in space by a scale of feet or inches; and in each case the scale coincides with the thing measured — mentally in the one; visibly in the other. So that in this simplest, and perhaps earliest case of quantitative prevision, the phenomena are not only thrust daily upon men's notice, but Nature is, as it were, perpetually repeating that process of measurement by observing which the prevision is effected. And thus there may be significance in the remark which some have made, that alike in Hebrew, Greek, and Latin, there is an affinity between the word meaning moon, and that meaning measure.

This fact, that in very early stages of social progress it is known that the moon goes through her changes in about thirty days, and that in about twelve moons the seasons return — this fact that chronological astronomy assumes a certain scientific character even before geometry does; while it is partly due to the circumstance that the astronomical divisions, day, month, and year, are ready made for us, is partly due to the further circumstances that agricultural and other operations were at first regulated astronomically, and that from the supposed divine nature of the heavenly bodies their motions determined the periodical religious festivals. As instances of the one we have the observation of the Egyptians, that the rising of the Nile corresponded with the heliacal rising of Sirius; the directions given by Hesiod for reaping and ploughing, according to the positions of the Pleiades; and his maxim that "fifty days after the turning of the sun is a seasonable time for beginning a voyage." As instances of the other, we have the naming of the days after the sun, moon, and planets; the early attempts among Eastern nations to regulate the calendar so that the gods might not be offended by the displacement of their sacrifices; and the fixing of the great annual festival of the Peruvians by the position of the sun. In all which facts we see that, at first, science was simply an appliance of religion and industry.

After the discoveries that a lunation occupies nearly thirty days, and that some twelve lunations occupy a year — discoveries of which there is no historical account, but which may be inferred as the earliest, from the fact that existing uncivilized races have made them — we come to the first known astronomical records, which are those of eclipses. The Chaldeans were able to predict these. "This they did, probably," says Dr. Whewell in his useful history, from which most of the materials we are about to use will be drawn, "by means of their cycle of 223 months, or about eighteen years; for at the end of this time, the eclipses of the moon begin to return, at the same intervals and in the same order as at the beginning." Now this method of calculating eclipses by means of a recurring cycle, — the *Saros* as they called it — is a more complex case of prevision by means of coincidence of measures. For by what observations must the Chaldeans have discovered this cycle? Obviously, as Delambre infers, by inspecting their registers; by comparing the successive intervals; by finding that some of the intervals were alike; by seeing that these equal intervals were eighteen years apart; by discovering that *all* the intervals that were eighteen years apart were equal; by ascertaining that the intervals formed a series which repeated itself, so that if one of the cycles of intervals were superposed on another the divisions would fit. This once perceived, and it manifestly became possible to use the cycle as a scale of time by which to measure out future periods. Seeing thus that the process of so predicting eclipses, is in essence the same as that of predicting the moon's monthly changes by observing the number of days after which they repeat — seeing that the two differ only in the extent and irregularity of the intervals, it is not difficult to understand how such an amount of knowledge should so early have been reached. And we shall be less surprised, on remembering that the only things involved in these previsions were *time* and *number;* and that the time was in a manner self-numbered.

Still, the ability to predict events recurring only after so long a period as eighteen years, implies a considerable advance in civilization — a considerable development of general knowledge; and we have now to inquire what progress in other sciences accompanied, and was necessary to, those astronomical previsions. In the first place, there must clearly have been a tolerably efficient system of calculation. Mere finger-counting, mere head-reckoning, even with the aid of a regular decimal notation, could not have sufficed for numbering the days in a year; much less the years, months, and days between eclipses. Consequently there must have been a mode of registering numbers; probably even a system of numerals. The earliest numerical records, if we may judge by the practices of the less civilized races now existing, were probably kept by notches cut on sticks, or strokes marked on walls; much as public-house scores are kept now. And there seems reason to believe that the first numerals used were simply groups of straight strokes, as some of the still-extant Roman ones are; leading us to suspect that these groups of strokes were used to represent groups of fingers, as the groups of fingers had been used to represent groups of objects — a supposition quite in conformity with the aboriginal system of picture writing and its subsequent modifications. Be this so or not, however, it is manifest that before the Chaldeans discovered their *Saros,* there must have been both a set of written symbols serving for an extensive numeration, and a familiarity with the simpler rules of arithmetic.

Not only must abstract mathematics have made some progress, but concrete mathematics also. It is scarcely possible that the buildings belonging to this era should have been laid out and erected without any knowledge of geometry. At any rate, there must have existed that elementary geometry which deals with direct measurement — with the apposition of lines; and it seems that only after the discovery of those simple proceedings, by which right angles are drawn, and relative positions fixed, could so regular an architecture be executed. In the case of the other division of concrete mathematics — mechanics, we have definite evidence of progress. We know that the lever and the inclined plane were employed during this period: implying that there was a qualitative prevision of their effects, though not a quantitative one. But we know more. We read of weights in the earliest records; and we find weights in ruins of the highest antiquity. Weights imply scales, of which we have also mention; and scales involve the primary theorem of mechanics in its least complicated form — involve not a qualitative but a quantitative prevision of mechanical effects. And here we may notice how mechanics, in common with the other exact sciences, took its rise from the simplest application of the idea of *equality.* For the mechanical proposition which the scales involve, is, that if a lever with *equal* arms, have *equal* weights suspended from them, the weights will remain at *equal* altitudes. And we may further notice, how, in this first step of rational mechanics, we see illustrated that truth awhile since referred to, that as magnitudes of linear extension are the only ones of which the equality is exactly ascertainable, the equalities of other magnitudes have at the outset to be determined by means of them. For the equality of the weights which balance each other in scales, wholly depends upon the equality of the arms: we can know that the weights are equal only by proving that the arms are equal. And when by this means we have obtained a system of weights, — a set of equal units of force, then does a science of mechanics become possible. Whence, indeed, it follows, that rational mechanics could not possibly have any other starting-point than the scales.

Let us further remember, that during this same period there was a limited knowledge of chemistry. The many arts which we know to have been carried on must have been impossible without a generalized experience of the modes in which certain bodies affect each other under special conditions. In metallurgy, which was extensively practised, this is abundantly illustrated. And we even have evidence that in some cases the knowledge possessed was, in a sense, quantitative. For, as we find by analysis that the hard alloy of which the Egyptians made their cutting tools, was composed of copper and tin in fixed proportions, there must have been an established prevision that each an alloy was to be obtained only by mixing them in these proportions. It is true, this was but a simple empirical generalization; but so was the generalization respecting the recurrence of eclipses; so are the first generalizations of every science.

Respecting the simultaneous advance of the sciences during this early epoch, it only remains to remark that even the most complex of them must have made some progress — perhaps even a greater relative progress than any of the rest. For under what conditions only were the foregoing developments possible? There first required an established and organized social system. A long continued registry of eclipses; the building of palaces; the use of scales; the practice of metallurgy — alike imply a fixed and populous nation. The existence of such a nation not only presupposes laws, and some administration of justice, which we know existed, but it presupposes successful laws — laws conforming in some degree to the conditions of social stability — laws enacted because it was seen that the actions forbidden by them were dangerous to the State. We do not by any means say that all, or even the greater part, of the laws were of this nature; but we do say, that the fundamental ones were. It cannot be denied that the laws affecting life and property were such. It cannot be denied that, however little these were enforced between class and class, they were to a considerable extent enforced between members of the same class. It can scarcely be questioned, that the administration of them between members of the same class was seen by rulers to be necessary for keeping their subjects together. And knowing, as we do, that, other things equal, nations prosper in proportion to the justness of their arrangements, we may fairly infer that the very cause of the advance of these earliest nations out of aboriginal barbarism, was the greater recognition among them of the claims to life and property.

But supposition aside, it is clear that the habitual recognition of these claims in their laws, implied some prevision of social phenomena. Even thus early there was a certain amount of social science. Nay, it may even be shown that there was a vague recognition of that fundamental principle on which all the true social science is based — the equal rights of all to the free exercise of their faculties. That same idea of *equality,* which, as we have seen, underlies all other science, underlies also morals and sociology. The conception of justice, which is the primary one in morals; and the administration of justice, which is the vital condition of social existence: are impossible, without the recognition of a certain likeness in men's claims, in virtue of their common humanity. *Equity* literally means *equalness;* and if it be admitted that there were even the vaguest ideas of equity in these primitive eras, it must be admitted that there was some appreciation of the equalness of men's liberties to pursue the objects of life — some appreciation, therefore, of the essential principle of national equilibrium.

Thus in this initial stage of the positive sciences, before geometry had yet done more than evolve a few empirical rules — before mechanics had passed beyond its first theorem — before astronomy had advanced from its merely chronological phase into the geometrical; the most involved of the sciences had reached a certain degree of development — a development without which no progress in other sciences was possible.

Only noting as we pass, how, thus early, we may see that the progress of exact science was not only towards an increasing number of previsions, but towards previsions more accurately quantitative — how, in astronomy, the recurring period of the moon's motions was by and by more correctly ascertained to be nineteen years, or two hundred and thirty-five lunations; how Callipus further corrected this Metonic cycle, by leaving out a day at the end of every seventy-six years; and how these successive advances implied a longer continued registry of observations, and the co-ordination of a greater number of facts — let us go on to inquire how geometrical astronomy took its rise.

The first astronomical instrument was the gnomon. This was not only early in use in the East, but it was found also among the Mexicans; the sole astronomical observations of the Peruvians were made by it; and we read that 1100 B. C., the Chinese found that, at a certain place, the length of the sun's shadow, at the summer solstice, was to the height of the gnomon, as one and a half to eight. Here again it is observable, not only that the instrument is found ready made, but that Nature is perpetually performing the process of measurement. Any fixed, erect object — a column, a dead palm, a pole, the angle of a building — serves for a gnomon; and it needs but to notice the changing position of the shadow it daily throws, to make the first step in geometrical astronomy. How small this first step was, may be seen in the fact that the only things ascertained at the outset were the periods of the summer and winter solstices, which corresponded with the least and greatest lengths of the mid-day shadow; and to fix which, it was needful merely to mark the point to which each day's shadow reached.

And now let it not be overlooked that in the observing at what time during the next year this extreme limit of the shadow was again reached, and in the inference that the sun had then arrived at the same turning point in his annual course, we have one of the simplest instances of that combined use of *equal magnitudes* and *equal relations,* by which all exact science, all quantitative prevision, is reached. For the relation observed was between the length of the sun's shadow and his position in the heavens; and the inference drawn was that when, next year, the extremity of his shadow came to the same point, he occupied the same place. That is, the ideas involved were, the equality of the shadows, and the equality of the relations between shadow and sun in successive years. As in the case of the scales, the equality of relations here recognized is of the simplest order. It is not as those habitually dealt with in the higher kinds of scientific reasoning, which answer to the general type — the relation between two and three equals the relation between six and nine; but it follows the type — the relation between two and three, equals the relation between two and three; it is a case of not simply *equal* relations, but *coinciding* relations. And here, indeed, we may see beautifully illustrated how the idea of equal relations takes its rise after the same manner that that of equal magnitude does. As already shown, the idea of equal magnitudes arose from the observed coincidence of two lengths placed together; and in this case we have not only two coincident lengths of shadows, but two coincident relations between sun and shadows.

From the use of the gnomon there naturally grew up the

conception of angular measurements; and with the advance of geometrical conceptions there came the hemisphere of Berosus, the equinoctial armil, the solstitial armil, and the quadrant of Ptolemy — all of them employing shadows as indices of the sun's position, but in combination with angular divisions. It is obviously out of the question for us here to trace these details of progress. It must suffice to remark that in all of them we may see that notion of equality of relations of a more complex kind, which is heat illustrated in the astrolabe, an instrument which consisted "of circular rims, moveable one within the other, or about poles, and contained circles which were to be brought into the position of the ecliptic, and of a plane passing through the sun and the poles of the ecliptic" — an instrument, therefore, which represented, as by a model, the relative positions of certain imaginary lines and planes in the heavens; which was adjusted by putting these representative lines and planes into parallelism and coincidence with the celestial ones; and which depended for its use upon the perception that the relations between these representative lines and planes were *equal* to the relations between those represented.

Were there space, we might go on to point out how the conception of the heavens as a revolving hollow sphere, the discovery of the globular form of the earth, the explanation of the moon's phases, and indeed all the successive steps taken, involved this same mental process. But we must content ourselves with referring to the theory of eccentrics and epicycles, as a further marked illustration of it. As first suggested, and as proved by Hipparchus to afford an explanation of the leading irregularities in the celestial motions, this theory involved the perception that the progressions, retrogressions, and variations of velocity seen in the heavenly bodies, might be reconciled with their assumed uniform movement in circles, by supposing that the earth was not in the centre of their orbits; or by supposing that they revolved in circles whose centres revolved round the earth; or by both. The discovery that this would account for the appearances, was the discovery that in certain geometrical diagrams the relations were such, that the uniform motion of a point would, when looked at from a particular position, present analogous irregularities; and the calculations of Hipparchus involved the belief that the relations subsisting among these geometrical curves were *equal* to the relations subsisting among the celestial orbits.

Leaving here these details of astronomical progress, and the philosophy of it, let us observe how the relatively concrete science of geometrical astronomy, having been thus far helped forward by the development of geometry in general, reacted upon geometry, caused it also to advance, and was again assisted by it. Hipparchus, before making his solar and lunar tables, had to discover rules for calculating the relations between the sides and angles of triangles — *trigonometry* a subdivision of pure mathematics. Further, the reduction of the doctrine of the sphere to the quantitative form needed for astronomical purposes, required the formation of a *spherical trigonometry,* which was also achieved by Hipparchus. Thus both plane and spherical trigonometry, which are parts of the highly abstract and simple science of extension, remained undeveloped until the less abstract and more complex science of the celestial motions had need of them. The fact admitted by M. Comte, that since Descartes the progress of the abstract division of mathematics has been determined by that of the concrete division, is paralleled by the still more significant fact that even thus early the progress of mathematics was determined by that of astronomy.

And here, indeed, we may see exemplified the truth, which the subsequent history of science frequently illustrates, that before any more abstract division makes a further advance, some more concrete division must suggest the necessity for that advance — must present the new order of questions to be solved. Before astronomy presented Hipparchus with the problem of solar tables, there was nothing to raise the question of the relations between lines and angles; the subject-matter of trigonometry had not been conceived. And as there must be subject-matter before there can be investigation, it follows that the progress of the concrete divisions is as necessary to that of the abstract, as the progress of the abstract to that of the concrete.

Just incidentally noticing the circumstance that the epoch we are describing witnessed the evolution of algebra, a comparatively abstract division of mathematics, by the union of its less abstract divisions, geometry and arithmetic — a fact proved by the earliest extant samples of algebra, which are half algebraic, half geometric — we go on to observe that during the era in which mathematics and astronomy were thus advancing, rational mechanics made its second step; and something was done towards giving a quantitative form to hydrostatics, optics, and harmonics. In each case we shall see as before, how the idea of equality underlies all quantitative prevision; and in what simple forms this idea is first applied.

As already shown, the first theorem established in mechanics was, that equal weights suspended from a lever with equal arms would remain in equilibrium. Archimedes discovered that a lever with unequal arms was in equilibrium when one weight was to its arm as the other arm to its weight; that is — when the numerical relation between one weight and its arm was *equal* to the numerical relation between the other arm and its weight.

The first advance made in hydrostatics, which we also owe to Archimedes, was the discovery that fluids press *equally* in all directions; and from this followed the solution of the problem of floating bodies: namely, that they are in equilibrium when the upward and downward pressures are *equal*.

In optics, again, the Greeks found that the angle of incidence is *equal* to the angle of reflection; and their knowledge reached no further than to such simple deductions from this as their geometry sufficed for. In harmonics they ascertained the fact that three strings of *equal* lengths would yield the octave, fifth and fourth, when strained by weights having certain definite ratios; and they did not progress much beyond this. In the one of which cases we see geometry used in elucidation of the laws of light; and in the other, geometry and arithmetic made to measure the phenomena of sound.

Did space permit, it would be desirable here to describe the state of the less advanced sciences — to point out how, while a few had thus reached the first stages of quantitative prevision, the rest were progressing in qualitative prevision — how some small generalizations were made respecting evaporation, and heat, and electricity, and magnetism, which, empirical as they were, did not in that respect differ from the first generalizations of every science — how the Greek physicians had made advances in physiology and pathology, which, considering the great imperfection of our present knowledge, are by no means to be despised — how zoology had been so far systematized by Aristotle, as, to some extent, enabled him from the presence of certain organs to predict the presence of others — how in Aristotle's *Politics,* there is some progress towards a scientific conception of social phenomena, and sundry previsions respecting them — and how in the state of the Greek societies, as well as in the writings of Greek philosophers, we may recognise not only an increasing clearness in that conception of equity on which the social science is based, but also some appreciation of the fact that social stability depends upon the maintenance of equitable regulations. We might dwell at length upon the causes which retarded the development of some of the sciences, as for example, chemistry; showing that relative complexity had nothing to do with it — that the oxidation of a piece of iron is a simpler phenomenon than the recurrence of eclipses, and the discovery of carbonic acid less difficult than that of the precession of the equinoxes — but that the relatively slow advance of chemical knowledge was due, partly to the fact that its phenomena were not daily thrust on men's notice as those of astronomy were; partly to the fact that Nature does not habitually supply the means, and suggest the modes of investigation, as in the sciences dealing with time extension, and force; and partly to the fact that the great majority of the materials with which chemistry deals, instead of being ready to hand, are made known only by the arts in their slow growth; and partly to the fact that even when known, their chemical properties are not self-exhibited, but have to be sought out by experiment.

Merely indicating all these considerations, however, let us go on to contemplate the progress and mutual influence of the sciences in modern days; only parenthetically noticing how, on the revival of the scientific spirit, the successive stages achieved exhibit the dominance of the same law hitherto traced — how the primary idea in dynamics, a uniform force, was defined by Galileo to be a force which generates *equal* velocities in *equal* successive times — how the uniform action of gravity was first experimentally determined by showing that the time elapsing before a body thrown up, stopped, was *equal* to the time it took to fall — how the first fact in compound motion which Galileo ascertained was, that a body projected horizontally will have a uniform motion onwards and a uniformly accelerated motion downwards; that is, will describe *equal* horizontal spaces in *equal* times, compounded with *equal* vertical increments in *equal* times — how his discovery respecting the pendulum was, that its oscillations occupy *equal* intervals of time whatever their length — how the principle of virtual velocities which he established is, that in any machine the weights that balance each other, are reciprocally as their virtual velocities; that is, the relation of one set of weights to their velocities *equals* the relation of the other set of velocities to their weights; — and how thus his achievements consisted in showing the equalities of certain magnitudes and relations, whose equalities had not been previously recognised.

When mechanics had reached the point to which Galileo brought it — when the simple laws of force had been disentangled from the friction and atmospheric resistance by which all their earthly manifestations are disguised — when progressing knowledge of *physics* had given a due insight into these disturbing causes — when, by an effort of abstraction, it was perceived that all motion would be uniform and rectilinear unless interfered with by external forces — and when the various consequences of this perception had been worked out; then it became possible, by the union of geometry and mechanics, to initiate physical astronomy. Geometry and mechanics having diverged from a common root in men's sensible experiences; having, with occasional inosculations, been separately developed, the one partly in connexion with astronomy, the other solely by analyzing terrestrial movements; now join in the investigations of Newton to create a true theory of the celestial motions. And here, also, we have to notice the important fact that, in the very process of being brought jointly to bear upon astronomical problems, they are themselves raised to a higher phase of development. For it was in dealing with the questions raised by celestial dynamics that the then incipient infinitesimal calculus was unfolded by Newton and his continental successors; and it was from inquiries into the mechanics of the solar system that the general theorems of mechanics contained in the "Principia," — many of them of purely terrestrial application — took their rise. Thus, as in the case of Hipparchus, the presentation of a new order of concrete facts to be analyzed, led to the discovery of pew abstract facts; and these abstract facts having been laid hold of, gave means of access to endless groups of concrete facts before incapable of quantitative treatment.

Meanwhile, physics had been carrying further that progress

without which, as just shown, rational mechanics could not be disentangled. In hydrostatics, Stevinus had extended and applied the discovery of Archimedes. Tor ricelli had proved atmospheric pressure, "by showing that this pressure sustained different liquids at heights inversely proportional to their densities;" and Pascal "established the necessary diminution of this pressure at increasing heights in the atmosphere:" discoveries which in part reduced this branch of science to a quantitative form. Something had been done by Daniel Bernouilli towards the dynamics of fluids. The thermometer had been invented; and a number of small generalizations reached by it. Huyghens and Newton had made considerable progress in optics; Newton had approximately calculated the rate of transmission of sound; and the continental mathematicians had succeeded in determining some of the laws of sonorous vibrations. Magnetism and electricity had been considerably advanced by Gilbert. Chemistry had got as far as the mutual neutralization of acids and alkalies. And Leonardo da Vinci had advanced in geology to the conception of the deposition of marine strata as the origin of fossils. Our present purpose does not require that we should give particulars. All that it here concerns us to do is to illustrate the *consensus* subsisting in this stage of growth, and afterwards. Let as look at a few cases.

The theoretic law of the velocity of sound enunciated by Newton on purely mechanical considerations, was found wrong by one-sixth. The error remained unaccounted for until the time of Laplace, who, suspecting that the heat disengaged by the compression of the undulating strata of the air, gave additional elasticity, and so produced the difference, made the needful calculations and found he was right. Thus acoustics was arrested until thermology overtook and aided it. When Boyle and Marriot had discovered the relation between the density of gases and the pressures they are subject to; and when it thus became possible to calculate the rate of decreasing density in the upper parts of the atmosphere; it also became possible to make approximate tables of the atmospheric refraction of light. Thus optics, and with it astronomy, advanced with barology. After the discovery of atmospheric pressure had led to the invention of the air-pump by Otto Guericke; and after it had become known that evaporation increases in rapidity as atmospheric pressure decreases; it became possible for Leslie, by evaporation in a vacuum, to produce the greatest cold known; and so to extend our knowledge of thermology by showing that there is no zero within reach of our researches. When Fourier had determined the laws of conduction of heat, and when the Earth's temperature had been found to increase below the surface one degree in every forty yards, there were data for inferring the past condition of our globe; the vast period it has taken to cool down to its present state; and the immense age of the solar system — a purely astronomical consideration.

Chemistry having advanced sufficiently to supply the needful materials, and a physiological experiment having furnished the requisite hint, there came the discovery of galvanic electricity. Galvanism reacting on chemistry disclosed the metallic bases of the alkalies, and inaugurated the electro-chemical theory; in the hands of Oersted and Ampère it led to the laws of magnetic action; and by its aid Faraday has detected significant facts relative to the constitution of light. Brewster's discoveries respecting double refraction and dipolarization proved the essential truth of the classification of crystalline forms according to the number of axes, by showing that the molecular constitution depends upon the axes. In these and in numerous other cases, the mutual influence of the sciences has been quite independent of any supposed hierarchical order. Often, too, their inter-actions are more complex than as thus instanced — involve more sciences than two. One illustration of this must suffice. We quote it in full from the *History of the Inductive Sciences.* In Book XI., chap. II., on "The Progress of the Electrical Theory," Dr. Whewell writes: —

"Thus at that period, mathematics was behind experiment, and a problem was proposed, in which theoretical results were wanted for comparison with observation, but could not be accurately obtained; as was the case in astronomy also, till the time of the approximate solution of the problem of three bodies, and the consequent formation of the tables of the moon and planets, on the theory of universal gravitation. After some time, electrical theory was relieved from this reproach, mainly in consequence of the progress which astronomy had occasioned in pure mathematics. About 1801 there appeared in the *Bulletin des Sciences,* an exact solution of the problem of the distribution of electric fluid on a spheroid, obtained by Biot, by the application of the peculiar methods which Laplace had invented for the problem of the figure of the planets. And, in 1811, M. Poisson applied Laplace's artifices to the case of two spheres acting upon one another in contact, a case to which many of Coulomb's experiments were referrible; and the agreement of the results of theory and observation, thus extricated from Coulomb's numbers obtained above forty years previously, was very striking and convincing."

Not only do the sciences affect each other after this direct manner, but they affect each other indirectly. Where there is no dependence, there is yet analogy — *equality of relations;* and the discovery of the relations subsisting among one set of phenomena, constantly suggests a search for the same relations among another set. Thus the established fact that the force of gravitation varies inversely as the square of the distance, being recognized as a necessary characteristic of all influences proceeding from a centre, raised the suspicion that heat and light follow the same law; which proved to be the case — a suspicion and a confirmation which were repeated in respect to the electric and magnetic forces. Thus again the discovery of the polarization of light led to experiments which ended in the discovery of the polarization of heat — a discovery that could never have been made without the antecedent one. Thus, too, the known refrangibility of light and heat lately produced the inquiry whether sound also is not refrangible; which on trial it turns out to be.

In some cases, indeed, it is only by the aid of conceptions derived from one class of phenomena that hypotheses respecting other classes can be formed. The theory, at one time favoured, that evaporation is a solution of water in air, was an assumption that the relation between water and air is *like* the relation between salt and water; and could never have been conceived if the relation between salt and water had not been previously known. Similarly the received theory of evaporation — that it is a diffusion of the particles of the evaporating fluid in virtue of their atomic repulsion — could not have been entertained without a foregoing experience of magnetic and electric repulsions. So complete in recent days has become this *consensus* among the sciences, caused either by the natural entanglement of their phenomena, or by analogies in the relations of their phenomena, that scarcely any considerable discovery concerning one order of facts now takes place, without very shortly leading to discoveries concerning other orders.

To produce a tolerably complete conception of this process of scientific evolution, it would be needful to go back to the beginning, and trace in detail the growth of classifications and nomenclatures; and to show how, as subsidiary to science, they have acted upon it, and it has reacted upon them. We can only now remark that, on the one hand, classifications and nomenclatures have aided science by continually subdividing the subject-matter of research, and giving fixity and diffusion to the truths disclosed; and that on the other hand, they have caught from it that increasing quantitativeness, and that progress from considerations touching single phenomena to considerations touching the relations among many phenomena, which we have been describing.

Of this last influence a few illustrations must be given. In chemistry it is seen in the facts, that the dividing of matter into the four elements was ostensibly based upon the single property of weight; that the first truly chemical division into acid and alkaline bodies, grouped together bodies which had not simply one property in common, but in which one property was constantly related to many others; and that the classification now current, places together in groups *supporters of combustion, metallic and non-metallic bases, acids, salts,* &c., bodies which are often quite unlike in sensible qualities, but which are like in the majority of their *relations* to other bodies. In mineralogy again, the first classifications were based upon differences in aspect, texture, and other physical attributes. Berzelius made two attempts at a classification based solely on chemical constitution. That now current, recognises as far as possible the *relations* between physical and chemical characters. In botany the earliest classes formed were *trees, shrubs,* and *herbs:* magnitude being the basis of distinction. Dioscorides divided vegetables into *aromatic, alimentary, medicinal,* and *vinous:* a division of chemical character. Cæsalpinus classified them by the seeds, and seed-vessels, which he preferred because of the *relations* found to subsist between the character of the fructification and the general character of the other parts.

While the "natural system" since developed, carrying out the doctrine of Linnæus, that "natural orders must be formed by attention not to one or two, but to *all* the parts of plants," bases its divisions on like peculiarities which are found to be *constantly related* to the greatest number of other like peculiarities. And similarly in zoology, the successive classifications, from having been originally determined by external and often subordinate characters not indicative of the essential nature, have been gradually more and more determined by those internal and fundamental differences, which have uniform *relations* to the greatest number of other differences. Nor shall we be surprised at this analogy between the modes of progress of positive science and classification, when we bear in mind that both proceed by making generalizations; that both enable us to make previsions differing only in their precision; and that while the one deals with equal properties and relations, the other deals with properties and relations that approximate towards equality in variable degrees.

Without further argument, it will, we think, be sufficiently clear that the sciences are none of them separately evolved — are none of them independent either logically or historically; but that all of them have, in a greater or less degree, required aid and reciprocated it. Indeed, it needs but to throw aside theses, and contemplate the mixed character of surrounding phenomena, to at once see that these notions of division and succession in the kinds of knowledge are none of them actually true, but are simple scientific fictions good, if regarded merely as aids to study; bad, if regarded as representing realities in Nature. Consider them critically, and no facts whatever are presented to our senses uncombined with other facts — no facts whatever but are in some degree disguised by accompanying facts: disguised in such a manner that all must be partially understood before any one can be understood. If it be said, as by M. Comte, that gravitating force should be treated of before other forces, seeing that all things are subject to it, it may on like grounds be said that heat should be first dealt with; seeing that thermal forces are everywhere in action; that the ability of any portion of matter to manifest visible gravitative phenomena depends on its state of aggregation, which is determined by heat; that only by the aid of thermology can we explain those apparent exceptions to the gravitating tendency which are presented by steam and smoke, and so establish its universality, and that, indeed, the very existence of the solar system in a solid form is just as much a question of heat as it is one of gravitation.

Take other cases: — All phenomena recognised by the eyes, through which only are the data of exact science ascertainable, are complicated with optical phenomena; and cannot be exhaustively known until optical principles are known. The burning of a candle cannot be explained without involving chemistry, mechanics, thermology. Every wind that blows is determined by influences partly solar, partly lunar, partly hygrometric; and implies considerations of fluid equilibrium and physical geography The direction, dip, and variations of the magnetic needle, are facts half terrestrial, half celestial — are caused by earthly forces which have cycles of change corresponding with astronomical periods. The flowing of the gulf-stream and the annual migration of icebergs towards the equator, depending as they do on the balancing of the centripetal and centrifugal forces acting on the ocean, involve in their explanation the Earth's rotation and spheroidal form, the laws of hydrostatics, the relative densities of cold and warm water, and the doctrines of evaporation. It is no doubt true, as M. Comte says, that "our position in the solar system, and the motions, form, size, equilibrium of the mass of our world among the planets, must be known before we can understand the phenomena going on at its surface." But, fatally for his hypothesis, it is also true that we must understand a great part of the phenomena going on at its surface before we can know its position, &c., in the solar system. It is not simply that, as we have already shown, those geometrical and mechanical principles by which celestial appearances are explained, were first generalized from terrestrial experiences; but it is that the very obtainment of correct data, on which to base astronomical generalizations, implies advanced terrestrial physics.

Until after optics had made considerable advance, the Copernican system remained but a speculation. A single modern observation on a star has to undergo a careful analysis by the combined aid of various sciences — has to *be digested by the organism of the sciences;* which have severally to assimilate their respective parts of the observation, before the essential fact it contains is available for the further development of astronomy. It has to be corrected not only for nutation of the earth's axis and for precession of the equinoxes, but for aberration and for refraction; and the formation of the tables by which refraction is calculated, presupposes knowledge of the law of decreasing density in the upper atmospheric strata; of the law of decreasing temperature, and the influence of this on the density; and of hygrometric laws as also affecting density. So that, to get materials for further advance, astronomy requires not only the indirect aid of the sciences which have presided over the making of its improved instruments, but the direct aid of an advanced optics, of barology, of thermology, of hygrometry; and if we remember that these delicate observations are in some cases registered electrically, and that they are further corrected for the "personal equation" — the time elapsing between seeing and registering, which varies with different observers — we may even add electricity and psychology. If, then, so apparently simple a thing as ascertaining the position of a star is complicated with so many phenomena, it is clear that this notion of the independence of the sciences, or certain of them, will not hold.

Whether objectively independent or not, they cannot be subjectively so — they cannot have independence as presented to our consciousness; and this is the only kind of independence with which we are concerned. And here, before leaving these illustrations, and especially this last one, let us not omit to notice how clearly they exhibit that increasingly active *consensus* of the sciences which characterizes their advancing development. Besides finding that in these later times a discovery in one science commonly causes progress in others; besides finding that a great part of the questions with which modern science deals are so mixed as to require the co-operation of many sciences for their solution; we find in this last case that, to make a single good observation in the purest of the natural sciences, requires the combined assistance of half a dozen other sciences.

Perhaps the clearest comprehension of the interconnected growth of the sciences may be obtained by contemplating that of the arts, to which it is strictly analogous, and with which it is inseparably bound up. Most intelligent persons must have been, at one time or other, struck with the vast array of antecedents pre-supposed by one of our processes of manufacture. Let him trace the production of a printed cotton, and consider all that is implied by it. There are the many successive improvements through which the power-looms reached their present perfection; there is the steam-engine that drives them, having its long history from Papin downwards; there are the lathes in which its cylinder was bored, and the string of ancestral lathes from which those lathes proceeded; there is the steam-hammer under which its crank shaft was welded; there are the puddling-furnaces, the blast-furnaces, the coal-mines and the iron-mines needful for producing the raw material; there are the slowly improved appliances by which the factory was built, and lighted, and ventilated; there are the printing engine, and the die house, and the colour laboratory with its stock of materials from all parts of the world, implying cochineal-culture, logwood-cutting, indigo-growing; there are the implements used by the producers of cotton, the gins by which it is cleaned, the elaborate machines by which it is spun: there are the vessels in which cotton is imported, with the building-slips, the rope-yards, the sail-cloth factories, the anchor-forges, needful for making them; and besides all these directly necessary antecedents, each of them involving many others, there are the institutions which have developed the requisite intelligence, the printing and publishing arrangements which have spread the necessary information, the social organization which has rendered possible such a complex co-operation of agencies.

Further analysis would show that the many arts thus concerned in the economical production of a child's frock, have each of them been brought to its present efficiency by slow steps which the other arts have aided; and that from the beginning this reciprocity has been ever on the increase. It needs but on the one hand to consider how utterly impossible it is for the savage, even with ore and coal ready, to produce so simple a thing as an iron hatchet; and then to consider, on the other hand, that it would have been impracticable among ourselves, even a century ago, to raise the tubes of the Britannia bridge from lack of the hydraulic press; to at once see how mutually dependent are the arts, and how all must advance that each may advance. Well, the sciences are involved with each other in just the same manner. They are, in fact, inextricably woven into this same complex web of the arts; and are only conventionally independent of it. Originally the two were one. How to fix the religious festivals; when to sow: how to weigh commodities; and in what manner to measure ground; were the purely practical questions out of which arose astronomy, mechanics, geometry. Since then there has been a perpetual inosculation of the sciences and the arts. Science has been supplying art with truer generalizations and more completely quantitative previsions. Art has been supplying science with better materials and more perfect instruments. And all along the interdependence has been growing closer, not only between art and science, but among the arts themselves, and among the sciences themselves.

How completely the analogy holds throughout, becomes yet clearer when we recognise the fact that *the sciences are arts to each other.* If, as occurs in almost every case, the fact to be analyzed by any science, has first to be prepared — to be disentangled from disturbing facts by the afore discovered methods of other sciences; the other sciences so used, stand in the position of arts. If, in solving a dynamical problem, a parallelogram is drawn, of which the sides and diagonal represent forces, and by putting magnitudes of extension for magnitudes of force a measurable relation is established between quantities not else to be dealt with; it may be fairly said that geometry plays towards mechanics much the same part that the fire of the founder plays towards the metal he is going to cast. If, in analyzing the phenomena of the coloured rings surrounding the point of contact between two lenses, a Newton ascertains by calculation the amount of certain interposed spaces, far too minute for actual measurement; he employs the science of number for essentially the same purpose as that for which the watchmaker employs tools. If, before writing down his observation on a star, the astronomer has to separate from it all the errors resulting from atmospheric and optical laws, it is manifest that the refraction-tables, and logarithm-books, and formulæ, which he successively uses, serve him much as retorts, and filters, and cupels serve the assayer who wishes to separate the pure gold from all accompanying ingredients.

So close, indeed, is the relationship, that it is impossible to say where science begins and art ends. All the instruments of the natural philosopher are the products of art; the adjusting one of them for use is an art; there is art in making an observation with one of them; it requires art properly to treat the facts ascertained; nay, even the employing established generalizations to open the way to new generalizations, may be considered as art. In each of these cases previously organized knowledge becomes the implement by which new knowledge is got at: and whether that previously organized knowledge is embodied in a tangible apparatus or in a formula, matters not in so far as its essential relation to the new knowledge is concerned. If, as no one will deny, art is applied knowledge, then such portion of a scientific investigation as consists of applied knowledge is art. So that we may even say that as soon as any prevision in science passes out of its originally passive state, and is employed for reaching other previsions, it passes from theory into practice — becomes science in action — becomes art. And when we thus see how purely conventional is the ordinary distinction, how impossible it is to make any real separation — when we see not only that science and art were originally one; that the arts have perpetually assisted each other; that there has been a constant reciprocation of aid between the sciences and arts; but that the sciences act as arts to each other, and that the established part of each science becomes an art to the growing part — when we recognize the closeness of these associations, we shall the more clearly perceive that as the connexion of the arts with each other has been ever becoming more intimate; as the help given by sciences to arts and by arts to sciences, has been age by age increasing; so the interdependence of the sciences themselves has been ever growing greater, their mutual relations more involved, their *consensus* more active.

In here ending our sketch of the Genesis of Science, we are conscious of having done the subject but scant justice. Two difficulties have stood in our way: one, the having to touch on so many points in such small space; the other, the necessity of treating in serial arrangement a process which is not serial — a difficulty which must ever attend all attempts to delineate processes of development, whatever their special nature. Add to which, that to present in anything like completeness and proportion, even the outlines of so vast and complex a history, demands years of study. Nevertheless, we believe that the evidence which has been assigned suffices to substantiate the leading propositions with which we set out. Inquiry into the first stages of science confirms the conclusion which we drew from the analysis of science as now existing, that it is not distinct from common knowledge, but an outgrowth from it — an extension of the perception by means of the reason.

That which we further found by analysis to form the more specific characteristic of scientific previsions, as contrasted with the previsions of uncultured intelligence — their quantitativeness — we also see to have been the characteristic alike in the initial steps in science, and of all the steps succeeding them. The facts and admissions cited in disproof of the assertion that the sciences follow one another, both logically and historically, in the order of their decreasing generality, have been enforced by the sundry instances we have met with, in which the more general or abstract sciences have been advanced only at the instigation of the more special or concrete — instances serving to show that a more general science as much owes its progress to the presentation of new problems by a more special science, as the more special science owes its progress to the solutions which the more general science is thus led to attempt — instances therefore illustrating the position that scientific advance is as much from the special to the general as from the general to the special.

Quite in harmony with this position we find to be the admissions that the sciences are as branches of one trunk, and that they were at first cultivated simultaneously; and this harmony becomes the more marked on finding, as we have done, not only that the sciences have a common root, but that science in general has a common root with language, classification, reasoning, art; that throughout civilization these have advanced together, acting and reacting upon each other just as the separate sciences have done; and that thus the development of intelligence in all its divisions and subdivisions has conformed to this same law which we have shown that the sciences conform to. From all which we may perceive that the sciences can with no greater propriety be arranged in a succession, than language, classification, reasoning, art, and science, can be arranged in a succession; that, however needful a succession may be for the convenience of books and catalogues, it must be recognized merely as a convention; and that so far from its being the function of a philosophy of the sciences to establish a hierarchy, it is its function to show that the linear arrangements required for literary purposes, have none of them any basis either in Nature or History.

There is one further remark we must not omit — a remark touching the importance of the question that has been discussed. Unfortunately it commonly happens that topics of this abstract nature are slighted as of no practical moment; and, we doubt not, that many will think it of very little consequence what theory respecting the genesis of science may be entertained. But the value of truths is often great, in proportion as their generality is wide. Remote as they seem from practical application, the highest generalizations are not unfrequently the most potent in their effects, in virtue of their influence on all those subordinate generalizations which regulate practice. And it must be so here. Whenever established, a correct theory of the historical development of the sciences must have an immense effect upon education; and, through education, upon civilisation. Greatly as we differ from him in other respects, we agree with M. Comte in the belief that, rightly conducted, the education of the individual must have a certain correspondence with the evolution of the race.

No one can contemplate the facts we have cited in illustration of the early stages of science, without recognising the *necessity* of the processes through which those stages were reached — a necessity which, in respect to the leading truths, may likewise be traced in all after stages. This necessity, originating in the very nature of the phenomena to be analyzed and the faculties to be employed, more or less fully applies to the mind of the child as to that of the savage. We say more or less fully, because the correspondence is not special but general only. Were the *environment* the same in both cases, the correspondence would be complete. But though the surrounding material out of which science is to be organized, is, in many cases, the same to the juvenile mind and the aboriginal mind, it is not so throughout; as, for instance, in the case of chemistry, the phenomena of which are accessible to the one, but were inaccessible to the other. Hence, in proportion as the environment differs, the course of evolution must differ. After admitting sundry exceptions, however, there remains a substantial parallelism; and, if so, it becomes of great moment to ascertain what really has been the process of scientific evolution. The establishment of an erroneous theory must be disastrous in its educational results; while the establishment of a true one must eventually be fertile in school-reforms and consequent social benefits.

IV
THE PHYSIOLOGY OF LAUGHTER

Why do we smile when a child puts on a man's hat? or what induces us to laugh on reading that the corpulent Gibbon was unable to rise from his knees after making a tender declaration? The usual reply to such questions is, that laughter results from a perception of incongruity. Even were there not on this reply the obvious criticism that laughter often occurs from extreme pleasure or from mere vivacity, there would still remain the real problem — How comes a sense of the incongruous to be followed by these peculiar bodily actions? Some have alleged that laughter is due to the pleasure of a relative self-elevation, which we feel on seeing the humiliation of others. But this theory, whatever portion of truth it may contain, is, in the first place, open to the fatal objection, that there are various humiliations to others which produce in us anything but laughter; and, in the second place, it does not apply to the many instances in which no one's dignity is implicated: as when we laugh at a good pun. Moreover, like the other, it is merely a generalization of certain conditions to laughter; and not an explanation of the odd movements which occur under these conditions. Why, when greatly delighted, or impressed with certain unexpected contrasts of ideas, should there be a contraction of particular facial muscles, and particular muscles of the chest and abdomen? Such answer to this question as may be possible, can be rendered only by physiology.

Every child has made the attempt to hold the foot still while it is tickled, and has failed; and probably there is scarcely any one who has not vainly tried to avoid winking, when a hand has been suddenly passed before the eyes. These examples of muscular movements which occur independently of the will, or in spite of it, illustrate what physiologists call reflex-action; as likewise do sneezing and coughing. To this class of cases, in which involuntary motions are accompanied by sensations, has to be added another class of cases, in which involuntary motions are unaccompanied by sensations: — instance the pulsations of the heart; the contractions of the stomach during digestion. Further, the great mass of seemingly-voluntary acts in such creatures as insects, worms, molluscs, are considered by physiologists to be as purely automatic as is the dilatation or closure of the iris under variations in quantity of light; and similarly exemplify the law, that an impression on the end of an afferent nerve is conveyed to some ganglionic centre, and is thence usually reflected along an efferent nerve to one or more muscles which it causes to contract.

In a modified form this principle holds with voluntary acts. Nervous excitation always *tends* to beget muscular motion; and when it rises to a certain intensity, always does beget it. Not only in reflex actions, whether with or without sensation, do we see that special nerves, when raised to a state of tension, discharge themselves on special muscles with which they are indirectly connected; but those external actions through which we read the feelings of others, show us that under any considerable tension, the nervous system in general discharges itself on the muscular system in general: either with or without the guidance of the will. The shivering produced by cold, implies irregular muscular contractions, which, though at first only partly involuntary, become, when the cold is extreme, almost wholly involuntary. When you have severely burnt your finger, it is very difficult to preserve a dignified composure: contortion of face, or movement of limb, is pretty sure to follow. If a man receives good news with neither change of feature nor bodily motion, it is inferred that he is not much pleased, or that he has extraordinary self-control — either inference implying that joy almost universally produces contraction of the muscles; and so, alters the expression, or attitude, or both. And when we hear of the feats of strength which men have performed when their lives were at stake — when we read how, in the energy of despair, even paralytic patients have regained for a time the use of their limbs; we see still more clearly the relations between nervous and muscular excitements. It becomes manifest both that emotions and sensations tend to generate bodily movements, and that the movements are vehement in proportion as the emotions or sensations are intense.[7]

This, however, is not the sole direction in which nervous excitement expends itself. Viscera as well as muscles may receive the discharge. That the heart and blood-vessels (which, indeed, being all contractile, may in a restricted sense be classed with the muscular system) are quickly affected by pleasures and pains, we have daily proved to us. Every sensation of any acuteness accelerates the pulse; and how sensitive the heart is to emotions, is testified by the familiar expressions which use heart and feeling as convertible terms. Similarly with the digestive organs. Without detailing the various ways in which these may be influenced by our mental states, it suffices to mention the marked benefits derived by dyspeptics, as well as other invalids, from cheerful society, welcome news, change of scene, to show how pleasurable feeling stimulates the viscera in general into greater activity.

There is still another direction in which any excited portion of the nervous system may discharge itself; and a direction in which it usually does discharge itself when the excitement is not strong. It may pass on the stimulus to some other portion of the nervous system. This is what occurs in quiet thinking and feeling. The successive states which constitute consciousness, result from this. Sensations excite ideas and emotions; these in their turns arouse other ideas and emotions; and so, continuously. That is to say, the tension existing in particular nerves, or groups of nerves, when they yield us certain sensations, ideas, or emotions, generates an equivalent tension in some other nerves, or groups of nerves, with which there is a connexion: the flow of energy passing on, the one idea or feeling dies in producing the next.

Thus, then, while we are totally unable to comprehend how the excitement of certain nerves should generate feeling — while, in the production of consciousness by physical agents acting on physical structure, we come to an absolute mystery never to be solved; it is yet quite possible for us to know by observation what are the successive forms which this absolute

7 For numerous illustrations see essay on "The Origin and Functions of Music."

mystery may take. We see that there are three channels along which nerves in a state of tension may discharge themselves; or rather, I should say, three classes of channels. They may pass on the excitement to other nerves that have no direct connexions with the bodily members, and may so cause other feelings and ideas; or they may pass on the excitement to one or more motor nerves, and so cause muscular contractions; or they may pass on the excitement to nerves which supply the viscera, and may so stimulate one or more of these.

For simplicity's sake, I have described these as alternative routes, one or other of which any current of nerve-force must take; thereby, as it may be thought, implying that such current will be exclusively confined to some one of them. But this is by no means the case. Rarely, if ever, does it happen that a state of nervous tension, present to consciousness as a feeling, expends itself in one direction only. Very generally it may be observed to expend itself in two; and it is probable that the discharge is never absolutely absent from any one of the three. There is, however, variety in the *proportions* in which the discharge is divided among these different channels under different circumstances. In a man whose fear impels him to run, the mental tension generated is only in part transformed into a muscular stimulus: there is a surplus which causes a rapid current of ideas. An agreeable state of feeling produced, say by praise, is not wholly used up in arousing the succeeding phase of the feeling, and the new ideas appropriate to it; but a certain portion overflows into the visceral nervous system, increasing the action of the heart, and probably facilitating digestion. And here we come upon a class of considerations and facts which open the way to a solution of our special problem.

For starting with the unquestionable truth, that at any moment the existing quantity of liberated nerve-force, which in an inscrutable way produces in us the state we call feeling, *must* expend itself in some direction — *must* generate an equivalent manifestation of force somewhere — it clearly follows that, if of the several channels it may take, one is wholly or partially closed, more must be taken by the others; or that if two are closed, the discharge along the remaining one must be more intense; and that, conversely, should anything determine an unusual efflux in one direction, there will be a diminished efflux in other directions.

Daily experience illustrates these conclusions. It is commonly remarked, that the suppression of external signs of feeling, makes feeling more intense. The deepest grief is silent grief. Why? Because the nervous excitement not discharged in muscular action, discharges itself in other nervous excitements — arouses more numerous and more remote associations of melancholy ideas, and so increases the mass of feelings. People who conceal their anger are habitually found to be more revengeful than those who explode in loud speech and vehement action. Why? Because, as before, the emotion is reflected back, accumulates, and intensifies. Similarly, men who, as proved by their powers of representation, have the keenest appreciation of the comic, are usually able to do and say the most ludicrous things with perfect gravity.

On the other hand, all are familiar with the truth that bodily activity deadens emotion. Under great irritation we get relief by walking about rapidly. Extreme effort in the bootless attempt to achieve a desired end, greatly diminishes the intensity of the desire. Those who are forced to exert themselves after misfortunes, do not suffer nearly so much as those who remain quiescent. If any one wishes to check intellectual excitement, he cannot choose a more efficient method than running till he is exhausted. Moreover, these cases, in which the production of feeling and thought is hindered by determining the nervous energy towards bodily movements, have their counterparts in the cases in which bodily movements are hindered by extra absorption of nervous energy in sudden thoughts and feelings. If, when walking along, there flashes on you an idea that creates great surprise, hope, or alarm, you stop; or if sitting cross-legged, swinging your pendent foot, the movement is at once arrested. From the viscera, too, intense mental action abstracts energy. Joy, disappointment, anxiety, or any moral perturbation rising to a great height, will destroy appetite; or if food has been taken, will arrest digestion; and even a purely intellectual activity, when extreme, will do the like.

Facts, then, fully bear out these *à priori* inferences, that the nervous excitement at any moment present to consciousness as feeling, must expend itself in some way or other; that of the three classes of channels open to it, it must take one, two, or more, according to circumstances; that the closure or obstruction of one, must increase the discharge through the others; and conversely, that if to answer some demand, the efflux of nervous energy in one direction is unusually great, there must be a corresponding decrease of the efflux in other directions. Setting out from these premises, let us now see what interpretation is to be put on the phenomena of laughter.

That laughter is a display of muscular excitement, and so illustrates the general law that feeling passing a certain pitch habitually vents itself in bodily action, scarcely needs pointing out. It perhaps needs pointing out, however, that strong feeling of almost any kind produces this result. It is not a sense of the ludicrous, only, which does it; nor are the various forms of joyous emotion the sole additional causes. We have, besides, the sardonic laughter and the hysterical laughter, which result from mental distress; to which must be added certain sensations, as tickling, and, according to Mr. Bain, cold, and some kinds of acute pain.

Strong feeling, mental or physical, being, then, the general cause of laughter, we have to note that the muscular actions constituting it are distinguished from most others by this, that they are purposeless. In general, bodily motions that are prompted by feelings are directed to special ends; as when we try to escape a danger, or struggle to secure a gratification. But the movements of chest and limbs which we make when laughing have no object. And now remark that these quasi-convulsive contractions of the muscles, having no object, but being results of an uncontrolled discharge of energy, we may see whence arise their special characters — how it happens that certain classes of muscles are affected first, and then certain other classes. For an overflow of nerve-force, undirected by any motive, will manifestly take first the most habitual routes; and if these do not suffice, will next overflow into the less habitual ones. Well, it is through the organs of speech that feeling passes into movement with the greatest frequency. The jaws, tongue, and lips are used not only to express strong irritation or gratification; but that very moderate flow of mental energy which accompanies ordinary conversation, finds its chief vent through this channel. Hence it happens that certain muscles round the mouth, small and easy to move, are the first to contract under pleasurable emotion. The class of muscles which, next after those of articulation, are most constantly set in action (or extra action, we should say) by feelings of all kinds, are those of respiration. Under pleasurable or painful sensations we breathe more rapidly: possibly as a consequence of the increased demand for oxygenated blood. The sensations that accompany exertion also bring on hard-breathing; which here more evidently responds to the physiological needs. And emotions, too, agreeable and disagreeable, both, at first, excite respiration; though the last subsequently depress it. That is to say, of the bodily muscles, the respiratory are more constantly implicated than any others in those various acts which our feelings impel us to; and, hence, when there occurs an undirected discharge of nervous energy into the muscular system, it happens that, if the quantity be considerable, it convulses not only certain of the articulatory and vocal muscles, but also those which expel air from the lungs.

Should the feeling to be expended be still greater in amount — too great to find vent in these classes of muscles — another class comes into play. The upper limbs are set in motion. Children frequently clap their hands in glee; by some adults the hands are rubbed together; and others, under still greater intensity of delight, slap their knees and sway their bodies backwards and forwards. Last of all, when the other channels for the escape of the surplus nerve-force have been filled to overflowing, a yet further and less-used group of muscles is spasmodically affected: the head is thrown back and the spine bent inwards — there is a slight degree of what medical men call opisthotonos. Thus, then, without contending that the phenomena of laughter in all their details are to be so accounted for, we see that in their *ensemble* they conform to these general principles: — that feeling excites to muscular action; that when the muscular action is unguided by a purpose, the muscles first affected are those which feeling most habitually stimulates; and that as the feeling to be expended increases in quantity, it excites an increasing number of muscles, in a succession determined by the relative frequency with which they respond to the regulated dictates of feeling.

There still, however, remains the question with which we set out. The explanation here given applies only to the laughter produced by acute pleasure or pain: it does not apply to the laughter that follows certain perceptions of incongruity. It is an insufficient explanation that in these cases, laughter is a result of the pleasure we take in escaping from the restraint of grave feelings. That this is a part-cause is true. Doubtless very often, as Mr. Bain says, "it is the coerced form of seriousness and solemnity without the reality that gives us that stiff position from which a contact with triviality or vulgarity relieves us, to our uproarious delight." And in so far as mirth is caused by the gush of agreeable feeling that follows the cessation of mental strain, it further illustrates the general principle above set forth. But no explanation is thus afforded of the mirth which ensues when the short silence between the *andante* and *allegro* in one of Beethoven's symphonies, is broken by a loud sneeze. In this, and hosts of like cases, the mental tension is not coerced but spontaneous — not disagreeable but agreeable; and the coming impressions to which the attention is directed, promise a gratification that few, if any, desire to escape. Hence, when the unlucky sneeze occurs, it cannot be that the laughter of the audience is due simply to the release from an irksome attitude of mind: some other cause must be sought.

This cause we shall arrive at by carrying our analysis a step further. We have but to consider the quantity of feeling that exists under such circumstances, and then to ask what are the conditions that determine the direction of its discharge, to at once reach a solution. Take a case. You are sitting in a theatre, absorbed in the progress of an interesting drama. Some climax has been reached which has aroused your sympathies — say, a reconciliation between the hero and heroine, after long and painful misunderstanding. The feelings excited by this scene are not of a kind from which you seek relief; but are, on the contrary, a grateful relief from the painful feelings with which you have witnessed the previous estrangement. Moreover, the sentiments these fictitious personages have for the moment inspired you with, are not such as would lead you to rejoice in any indignity offered to them; but rather, such as would make you resent the indignity. And now, while you are contemplating the reconciliation with a pleasurable sympathy, there appears from behind the scenes a tame kid, which, having stared round at the audience, walks up to the lovers and sniffs at them. You cannot help joining in the roar which greets this *contretemps*. Inexplicable as is this irresistible burst on the hypothesis of a pleasure in escaping from mental restraint; or on the hypothesis of a pleasure from relative increase of self-importance, when witnessing the humiliation of others; it is readily explicable if we consider what, in such a case, must become of the feeling that existed at the moment the incongruity arose. A large mass of emotion had been produced; or, to speak in physiological language, a large portion of the nervous system was in a state of tension. There was also great expectation with respect to the further evolution of the scene — a quantity of vague, nascent thought and emotion, into which the existing quantity of thought and emotion was about to pass.

Had there been no interruption, the body of new ideas and feelings next excited, would have sufficed to absorb the whole of the liberated nervous energy. But now, this large amount of nervous energy, instead of being allowed to expend itself in producing an equivalent amount of the new thoughts and emotions which were nascent, is suddenly checked in its flow. The channels along which the discharge was about to take place, are closed. The new channel opened — that afforded by the appearance and proceedings of the kid — is a small one; the ideas and feelings suggested are not numerous and massive enough to carry off the nervous energy to be expended. The excess must therefore discharge itself in some other direction; and in the way already explained, there results an efflux through the motor nerves to various classes of the muscles, producing the half-convulsive actions we term laughter.

This explanation is in harmony with the fact, that when, among several persons who witness the same ludicrous occurrence, there are some who do not laugh; it is because there has arisen in them an emotion not participated in by the rest, and which is sufficiently massive to absorb all the nascent excitement. Among the spectators of an awkward tumble, those who preserve their gravity are those in whom there is excited a degree of sympathy with the sufferer, sufficiently great to serve as an outlet for the feeling which the occurrence had turned out of its previous course. Sometimes anger carries off the arrested current; and so prevents laughter. An instance of this was lately furnished me by a friend who had been witnessing the feats at Franconi's. A tremendous leap had just been made by an acrobat over a number of horses. The clown, seemingly envious of this success, made ostentatious preparation for doing the like; and then, taking the preliminary run with immense energy, stopped short on reaching the first horse, and pretended to wipe some dust from its haunches. In the majority of the spectators, merriment was excited; but in my friend, wound up by the expectation of the coming leap to a state of great nervous tension, the effect of the baulk was to produce indignation. Experience thus proves what the theory implies: namely, that the discharge of arrested feelings into the muscular system, takes place only in the absence of other adequate channels — does not take place if there arise other feelings equal in amount to those arrested.

Evidence still more conclusive is at hand. If we contrast the incongruities which produce laughter with those which do not, we at once see that in the non-ludicrous ones the unexpected state of feeling aroused, though wholly different in kind, is not less in quantity or intensity. Among incongruities that may excite anything but a laugh, Sir. Bain instances — "A decrepit man under a heavy burden, five loaves and two fishes among a multitude, and all unfitness and gross disproportion; an instrument out of tune, a fly in ointment, snow in May, Archimedes studying geometry in a siege, and all discordant things; a wolf in sheep's clothing, a breach of bargain, and falsehood in general; the multitude taking the law in their own hands, and everything of the nature of disorder; a corpse at a feast, parental cruelty, filial ingratitude, and whatever is unnatural; the entire catalogue of the vanities given by Solomon, are all incongruous, but they cause feelings of pain, anger, sadness, loathing, rather than mirth." Now in these cases, where the totally unlike state of consciousness suddenly produced, is not inferior in mass to the preceding one, the conditions to laughter are not fulfilled. As above shown, laughter naturally results only when consciousness is unawares transferred from great things to small — only when there is what we call a *descending* incongruity.

And now observe, finally, the fact, alike inferable *à priori* and illustrated in experience, that an *ascending* incongruity not only fails to cause laughter, but works on the muscular system an effect of exactly the reverse kind. When after something very insignificant there arises without anticipation something very great, the emotion we call wonder results; and this emotion is accompanied not by an excitement of the muscles, but by a relaxation of them. In children and country people, that falling of the jaw which occurs on witnessing something that is imposing and unexpected, exemplifies this effect. Persons who have been wonder-struck at the production of very striking results by a seemingly inadequate cause, are frequently described as unconsciously dropping the things they held in their hands. Such are just the effects to be anticipated. After an average state of consciousness, absorbing but a small quantity of nervous energy, is aroused without the slightest notice, a strong emotion of awe, terror, or admiration; joined with the astonishment due to an apparent want of adequate causation. This new state of consciousness demands far more nervous energy than that which it has suddenly replaced; and this increased absorption of nervous energy in mental changes, involves a temporary diminution of the outflow in other directions: whence the pendent jaw and the relaxing grasp.

One further observation is worth making. Among the several sets of channels into which surplus feeling might be discharged, was named the nervous system of the viscera. The sudden overflow of an arrested mental excitement, which, as we have seen, results from a descending incongruity, must doubtless stimulate not only the muscular system, as we see it does, but also the internal organs; the heart and stomach must come in for a share of the discharge. And thus there seems to be a good physiological basis for the popular notion that mirth-creating excitement facilitates digestion.

Though in doing so I go beyond the boundaries of the immediate topic, I may fitly point out that the method of inquiry here followed, is one which enables us to understand various phenomena besides those of laughter. To show the importance of pursuing it, I will indicate the explanation it furnishes of another familiar class of facts.

All know how generally a large amount of emotion disturbs the action of the intellect, and interferes with the power of expression. A speech delivered with great facility to tables and chairs, is by no means so easily delivered to an audience. Every schoolboy can testify that his trepidation, when standing before a master, has often disabled him from repeating a lesson which he had duly learnt. In explanation of this we commonly say that the attention is distracted — that the proper train of ideas is broken by the intrusion of ideas that are irrelevant. But the question is, in what manner does unusual emotion produce this effect; and we are here supplied with a tolerably obvious answer. The repetition of a lesson, or set speech previously thought out, implies the flow of a very moderate amount of nervous excitement through a comparatively narrow channel. The thing to be done is simply to call up in succession certain previously-arranged ideas — a process in which no great amount of mental energy is expended. Hence, when there is a large quantity of emotion, which must be discharged in some direction or other; and when, as usually happens, the restricted series of intellectual actions to be gone through, does not suffice to carry it off; there result discharges along other channels besides the one prescribed: there are aroused various ideas foreign to the train of thought to be pursued; and these tend to exclude from consciousness those which should occupy it.

And now observe the meaning of those bodily actions spontaneously set up under these circumstances. The school-boy saying his lesson, commonly has his fingers actively engaged — perhaps in twisting about a broken pen, or perhaps squeezing the angle of his jacket; and if told to keep his hands still, he soon again falls into the same or a similar trick. Many anecdotes are current of public speakers having incurable automatic actions of this class: barristers who perpetually wound and unwound pieces of tape; members of parliament ever putting on and taking off their spectacles. So long as such movements are unconscious, they facilitate the mental actions. At least this seems a fair inference from the fact that confusion frequently results from putting a stop to them: witness the case narrated by Sir Walter Scott of his school-fellow, who became unable to say his lesson after the removal of the waistcoat-button that he habitually fingered while in class. But why do they facilitate the mental actions? Clearly because they draw off a portion of the surplus nervous excitement. If, as above explained, the quantity of mental energy generated is greater than can find vent along the narrow channel of thought that is open to it; and if, in consequence, it is apt to produce confusion by rushing into other channels of thought; then by allowing it an exit through the motor nerves into the muscular system, the pressure is diminished, and irrelevant ideas are less likely to intrude on consciousness.

This further illustration will, I think, justify the position that something may be achieved by pursuing in other cases this method of psychological inquiry. A complete explanation of the phenomena, requires us to trace out *all* the consequences of any given state of consciousness; and we cannot do this without studying the effects, bodily and mental, as varying in quantity at each other's expense. We should probably learn much if we in every case asked — Where is all the nervous energy gone?

V
THE ORIGIN AND FUNCTION OF MUSIC

When Carlo, standing, chained to his kennel, seen his master in the distance, a slight motion of the tail indicates his but faint hope that he is about to be let out. A much more decided wagging of the tail, passing by-and-by into lateral undulations of the body, follows his master's nearer approach. When hands are laid on his collar, and he knows that he is really to have an outing, his jumping and wriggling are such that it is by no means easy to loose his fastenings. And when he finds himself actually free, his joy expends itself in bounds, in pirouettes, and in scourings hither and thither at the top of his speed. Puss, too, by erecting her tail, and by every time raising her back to meet the caressing hand of her mistress, similarly expresses her gratification by certain muscular actions; as likewise do the parrot by awkward dancing on his perch, and the canary by hopping and fluttering about his cage with unwonted rapidity. Under emotions of an opposite kind, animals equally display muscular excitement. The enraged lion lashes his sides with his tail, knits his brows, protrudes his claws. The cat sets up her back; the dog retracts his upper lip; the horse throws back his ears. And in the struggles of creatures in pain, we see that the like relation holds between excitement of the muscles and excitement of the nerves of sensation.

In ourselves, distinguished from lower creatures as we are by feelings alike more powerful and more varied, parallel facts are at once more conspicuous and more numerous. We may conveniently look at them in groups. We shall find that pleasurable sensations and painful sensations, pleasurable emotions and painful emotions, all tend to produce active demonstrations in proportion to their intensity.

In children, and even in adults who are not restrained by regard for appearances, a highly agreeable taste is followed by a smacking of the lips. An infant will laugh and bound in its nurse's arms at the sight of a brilliant colour or the hearing of a new sound. People are apt to beat time with head or feet to music which particularly pleases them. In a sensitive person an agreeable perfume will produce a smile; and smiles will be seen on the faces of a crowd gazing at some splendid burst of fireworks. Even the pleasant sensation of warmth felt on getting to the fireside out of a winter's storm, will similarly express itself in the face.

Painful sensations, being mostly far more intense than pleasurable ones, cause muscular actions of a much more decided kind. A sudden twinge produces a convulsive start of the whole body. A pain less violent, but continuous, is accompanied by a knitting of the brows, a setting of the teeth or biting of the lip, and a contraction of the features generally. Under a persistent pain of a severer kind, other muscular actions are added: the body is swayed to and fro; the hands clench anything they can lay hold of; and should the agony rise still higher, the sufferer rolls about on the floor almost convulsed.

Though more varied, the natural language of the pleasurable emotions comes within the same generalization. A smile, which is the commonest expression of gratified feeling, is a contraction of certain facial muscles; and when the smile broadens into a laugh, we see a more violent and more general muscular excitement produced by an intenser gratification. Rubbing together of the hands, and that other motion which Dickens somewhere describes as "washing with impalpable soap in invisible water," have like implications. Children may often be seen to "jump for joy." Even in adults of excitable temperament, an action approaching to it is sometimes witnessed. And dancing has all the world through been regarded as natural to an elevated state of mind. Many of the special emotions show themselves in special muscular actions. The gratification resulting from success, raises the head and gives firmness to the gait. A hearty grasp of the hand is currently taken as indicative of friendship. Under a gush of affection the mother clasps her child to her breast, feeling as though she could squeeze it to death. And so in sundry other cases. Even in that brightening of the eye with which good news is received we may trace the same truth; for this appearance of greater brilliancy is due to an extra contraction of the muscle which raises the eyelid, and so allows more light to fall upon, and be reflected from, the wet surface of the eyeball.

The bodily indications of painful emotions are equally numerous, and still more vehement. Discontent is shown by raised eyebrows and wrinkled forehead; disgust by a curl of the lip; offence by a pout. The impatient man beats a tattoo with his fingers on the table, swings his pendent leg with increasing rapidity, gives needless pokings to the fire, and presently paces with hasty strides about the room. In great grief there is wringing of the hands, and even tearing of the hair. An angry child stamps, or rolls on its back and kicks its heels in the air; and in manhood, anger, first showing itself in frowns, in distended nostrils, in compressed lips, goes on to produce grinding of the teeth, clenching of the fingers, blows of the fist on the table, and perhaps ends in a violent attack on the offending person, or in throwing about and breaking the furniture. From that pursing of the mouth indicative of slight displeasure, up to the frantic struggles of the maniac, we shall find that mental irritation tends to vent itself in bodily activity.

All feelings, then — sensations or emotions, pleasurable or painful — have this common characteristic, that they are muscular stimuli. Not forgetting the few apparently exceptional cases in which emotions exceeding a certain intensity produce prostration, we may set it down as a general law that, alike in man and animals, there is a direct connection between feeling and motion; the last growing more vehement as the first grows more intense. Were it allowable here to treat the matter scientifically, we might trace this general law down to the principle known among physiologists as that of *reflex action.*[8] Without doing this, however, the above numerous instances justify the generalization, that mental excitement of all kinds ends in excitement of the muscles; and that the two preserve a more or less constant ratio to each other.

"But what has all this to do with *The Origin and Function of Music?*" asks the reader. Very much, as we shall presently see. All music is originally vocal. All vocal sounds are produced by the agency of certain muscles. These muscles, in common with

---

8 Those who seek information on this point may find it in an interesting tract by Mr. Alexander Bain, on *Animal Instinct and Intelligence.*

those of the body at large, are excited to contraction by pleasurable and painful feelings. And therefore it is that feelings demonstrate themselves in sounds as well as in movements. Therefore it is that Carlo barks as well as leaps when he is let out — that puss purrs as well as erects her tail — that the canary chirps as well as flutters. Therefore it is that the angry lion roars while he lashes his sides, and the dog growls while he retracts his lip. Therefore it is that the maimed animal not only struggles, but howls. And it is from this cause that in human beings bodily suffering expresses itself not only in contortions, but in shrieks and groans — that in anger, and fear, and grief, the gesticulations are accompanied by shouts and screams — that delightful sensations are followed by exclamations — and that we hear screams of joy and shouts of exultation.

We have here, then, a principle underlying all vocal phenomena; including those of vocal music, and by consequence those of music in general. The muscles that move the chest, larynx, and vocal chords, contracting like other muscles in proportion to the intensity of the feelings; every different contraction of these muscles involving, as it does, a different adjustment of the vocal organs; every different adjustment of the vocal organs causing a change in the sound emitted; — it follows that variations of voice are the physiological results of variations of feeling; it follows that each inflection or modulation is the natural outcome of some passing emotion or sensation; and it follows that the explanation of all kinds of vocal expression, must be sought in this general relation between mental and muscular excitements. Let us, then, see whether we cannot thus account for the chief peculiarities in the utterance of the feelings: grouping these peculiarities under the heads of *loudness, quality, or timbre, pitch, intervals,* and *rate of variation.*

Between the lungs and the organs of voice, there is much the same relation as between the bellows of an organ and its pipes. And as the loudness of the sound given out by an organ-pipe increases with the strength of the blast from the bellows; so, other things equal, the loudness of a vocal sound increases with the strength of the blast from the lungs. But the expulsion of air from the lungs is effected by certain muscles of the chest and abdomen. The force with which these muscles contract, is proportionate to the intensity of the feeling experienced. Hence, *à priori,* loud sounds will be the habitual results of strong feelings. That they are so we have daily proof. The pain which, if moderate, can be borne silently, causes outcries if it becomes extreme. While a slight vexation makes a child whimper, a fit of passion calls forth a howl that disturbs the neighbourhood. When the voices in an adjacent room become unusually audible, we infer anger, or surprise, or joy. Loudness of applause is significant of great approbation; and with uproarious mirth we associate the idea of high enjoyment. Commencing with the silence of apathy, we find that the utterances grow louder as the sensations or emotions, whether pleasurable or painful, grow stronger.

That different *qualities* of voice accompany different mental states, and that under states of excitement the tones are more sonorous than usual, is another general fact admitting of a parallel explanation. The sounds of common conversation have but little resonance; those of strong feeling have much more. Under rising ill temper the voice acquires a metallic ring. In accordance with her constant mood, the ordinary speech of a virago has a piercing quality quite opposite to that softness indicative of placidity. A ringing laugh marks an especially joyous temperament. Grief unburdening itself uses tones approaching in *timbre* to those of chanting: and in his most pathetic passages an eloquent speaker similarly falls into tones more vibratory than those common to him. Now any one may readily convince himself that resonant vocal sounds can be produced only by a certain muscular effort additional to that ordinarily needed. If after uttering a word in his speaking voice, the reader, without changing the pitch or the loudness, will *sing* this word, he will perceive that before he can sing it, he has to alter the adjustment of the vocal organs; to do which a certain force must be used; and by putting his fingers on that external prominence marking the top of the larynx, he will have further evidence that to produce a sonorous tone the organs must be drawn out of their usual position. Thus, then, the fact that the tones of excited feeling are more vibratory than those of common conversation, is another instance of the connexion between mental excitement and muscular excitement. The speaking voice, the recitative voice, and the singing voice, severally exemplify one general principle.

That the *pitch* of the voice varies according to the action of the vocal muscles, scarcely needs saying. All know that the middle notes, in which they converse, are made without any appreciable effort; and all know that to make either very high or very low notes requires a considerable effort. In either ascending or descending from the pitch of ordinary speech, we are conscious of an increasing muscular strain, which, at both extremes of the register, becomes positively painful. Hence it follows from our general principle, that while indifference or calmness will use the medium tones, the tones used during excitement will be either above or below them; and will rise higher and higher, or fall lower and lower, as the feelings grow stronger. This physiological deduction we also find to be in harmony with familiar facts. The habitual sufferer utters his complaints in a voice raised considerably above the natural key; and agonizing pain vents itself in either shrieks or groans — in very high or very low notes. Beginning at his talking pitch, the cry of the disappointed urchin grows more shrill as it grows louder. The "Oh!" of astonishment or delight, begins several notes below the middle voice, and descends still lower. Anger expresses it self in high tones, or else in "curses not loud but *deep.*" Deep tones, too, are always used in uttering strong reproaches. Such an exclamation as "Beware!" if made dramatically — that is, if made with a show of feeling — must be many notes lower than ordinary. Further, we have groans of disapprobation, groans of horror, groans of remorse. And extreme joy and fear are alike accompanied by shrill outcries.

Nearly allied to the subject of pitch, is that of *intervals;* and the explanation of them carries our argument a step further. While calm speech is comparatively monotonous, emotion makes use of fifths, octaves, and even wider intervals. Listen to any one narrating or repeating something in which he has no interest, and his voice will not wander more than two or three notes above or below his medium note, and that by small steps; but when he comes to some exciting event he will be heard not only to use the higher and lower notes of his register, but to go from one to the other by larger leaps. Being unable in print to imitate these traits of feeling, we feel some difficulty in fully realizing them to the reader. But we may suggest a few remembrances which will perhaps call to mind a sufficiency of others. If two men living in the same place, and frequently seeing one another, meet, say at a public assembly, any phrase with which one may be heard to accost the other — as "Hallo, are you here?" — will have an ordinary intonation. But if one of them, after long absence, has unexpectedly returned, the expression of surprise with which his friend may greet him — "Hallo! how came you here?" — will be uttered in much more strongly contrasted tones. The two syllables of the word "Hallo" will be, the one much higher and the other much lower than before; and the rest of the sentence will similarly ascend and descend by longer steps.

Again, if, supposing her to be in an adjoining room, the mistress of the house calls "Mary," the two syllables of the name will be spoken in an ascending interval of a third. If Mary does not reply, the call will be repeated probably in a descending fifth; implying the slightest shade of annoyance at Mary's inattention. Should Mary still make no answer, the increasing annoyance will show itself by the use of a descending octave on the next repetition of the call. And supposing the silence to continue, the lady, if not of a very even temper, will show her irritation at Mary's seemingly intentional negligence by finally calling her in tones still more widely contrasted — the first syllable being higher and the last lower than before.

Now, these and analogous facts, which the reader will readily accumulate, clearly conform to the law laid down. For to make large intervals requires more muscular action than to make small ones. But not only is the *extent* of vocal intervals thus explicable as due to the relation between nervous and muscular excitement, but also in some degree their *direction*, as ascending or descending. The middle notes being those which demand no appreciable effort of muscular adjustment; and the effort becoming greater as we either ascend or descend; it follows that a departure from the middle notes in either direction will mark increasing emotion; while a return towards the middle notes will mark decreasing emotion. Hence it happens that an enthusiastic person uttering such a sentence as — "It was the most splendid sight I ever saw!" will ascend to the first syllable of the word "splendid," and thence will descend: the word "splendid" marking the climax of the feeling produced by the recollection. Hence, again, it happens that, under some extreme vexation produced by another's stupidity, an irascible man, exclaiming — "What a confounded fool the fellow is!" will begin somewhat below his middle voice, and descending to the word "fool," which he will utter in one of his deepest notes, will then ascend again. And it may be remarked, that the word "fool" will not only be deeper and. louder than the rest, but will also have more emphasis of articulation — another mode in which muscular excitement is shown.

There is some danger, however, in giving instances like this; seeing that as the mode of rendering will vary according to the intensity of the feeling which the reader feigns to himself, the right cadence may not be hit upon. With single words there is less difficulty. Thus the "Indeed!" with which a surprising fact is received, mostly begins on the middle note of the voice, and rises with the second syllable; or, if disapprobation as well as astonishment is felt, the first syllable will be below the middle note, and the second lower still. Conversely, the word "Alas!" which marks not the rise of a paroxysm of grief, but its decline, is uttered in a cadence descending towards the middle note; or, if the first syllable is in the lower part of the register, the second ascends towards the middle note. In the "Heigh-ho!" expressive of mental and muscular prostration, we may see the same truth; and if the cadence appropriate to it be inverted the absurdity of the effect clearly shows how the meaning of intervals is dependent on the principle we have been illustrating.

The remaining characteristic of emotional speech which we have to notice is that of *variability of pitch.* It is scarcely possible here to convey adequate ideas of this more complex manifestation. We must be content with simply indicating some occasions on which it may be observed. On a meeting of friends, for instance — as when there arrives a party of much-wished-for visitors — the voices of all will be heard to undergo changes of pitch not only greater but much more numerous than usual. If a speaker at a public meeting is interrupted by some squabble among those he is addressing, his comparatively level tones will be in marked contrast with the rapidly changing one of the disputants. And among children, whose feelings are less under control than those of adults, this peculiarity is still more decided. During a scene of complaint and recrimination between two excitable little girls, the voices may be heard to run up and down the gamut several times in each sentence. In such cases we once more recognise the same law: for muscular excitement is shown not only in strength of contraction but also in the rapidity with which different muscular adjustments succeed each other.

Thus we find all the leading vocal phenomena to have a physiological basis. They are so many manifestations of the general law that feeling is a stimulus to muscular action — a law conformed to throughout the whole economy, not of man only, but of every sensitive creature — a law, therefore, which lies deep in the nature of animal organization. The expressiveness of these various modifications of voice is therefore innate. Each of us, from babyhood upwards, has been spontaneously making them, when under the various sensations and emotions by which they are produced. Having been conscious of each feeling at the same time that we heard ourselves make the consequent sound, we have acquired an established association of ideas between such sound and the feeling which caused it. When the like sound is made by another, we ascribe the like feeling to him; and by a further consequence we not only ascribe to him that feeling, but have a certain degree of it aroused in ourselves: for to become conscious of the feeling which another is experiencing, is to have that feeling awakened in our own consciousness, which is the same thing as experiencing the feeling. Thus these various modifications of voice become not only a language through which we understand the emotions of others, but also the means of exciting our sympathy with such emotions.

Have we not here, then, adequate data for a theory of music? These vocal peculiarities which indicate excited feeling, *are those which especially distinguish song from ordinary speech.* Every one of the alterations of voice which we have found to be a physiological result of pain or pleasure, *is carried to its greatest extreme in vocal music.* For instance, we saw that, in virtue of the general relation between mental and muscular excitement, one characteristic of passionate utterance is *loudness.* Well, its comparative loudness is one of the distinctive marks of song as contrasted with the speech of daily life; and further, the *forte* passages of an air are those intended to represent the climax of its emotion. We next saw that the tones in which emotion expresses itself, are, in conformity with this same law, of a more sonorous *timbre* than those of calm conversation. Here, too, song displays a still higher degree of the peculiarity; for the singing tone is the most resonant we can make. Again, it was shown that, from a like cause, mental excitement vents itself in the higher and lower notes of the register; using the middle notes but seldom. And it scarcely needs saying that vocal music is still more distinguished by its comparative neglect of the notes in which we talk, and its habitual use of those above or below them and, moreover, that its most passionate effects are commonly produced at the two extremities of its scale, but especially the upper one.

A yet further trait of strong feeling, similarly accounted for, was the employment of larger intervals than are employed in common converse. This trait, also, every ballad and *aria* carries to an extent beyond that heard in the spontaneous utterances of emotion: add to which, that the direction of these intervals, which, as diverging from or converging towards the medium tones, we found to be physiologically expressive of increasing or decreasing emotion, may be observed to have in music like meanings. Once more, it was pointed out that not only extreme but also rapid variations of pitch, are characteristic of mental excitement; and once more we see in the quick changes of every melody, that song carries the characteristic as far, if not farther. Thus, in respect alike of *loudness, timbre, pitch, intervals,* and *rate of variation,* song employs and exaggerates the natural language of the emotions; — it arises from a systematic combination of those vocal peculiarities which are the physiological effects of acute pleasure and pain.

Besides these chief characteristics of song as distinguished from common speech, there are sundry minor ones similarly explicable as due to the relation between mental and muscular excitement; and before proceeding further these should be briefly noticed. Thus, certain passions, and perhaps all passions when pushed to an extreme, produce (probably through their influence over the action of the heart) an effect the reverse of that which has been described: they cause a physical prostration, one symptom of which is a general relaxation of the muscles, and a consequent trembling. We have the trembling of anger, of fear, of hope, of joy; and the vocal muscles being implicated with the rest, the voice too becomes tremulous. Now, in singing, this tremulousness of voice is very effectively used by some vocalists in highly pathetic passages; sometimes, indeed, because of its effectiveness, too much used by them — as by Tamberlik, for instance.

Again, there is a mode of musical execution known as the *staccato,* appropriate to energetic passages — to passages expressive of exhilaration, of resolution, of confidence. The action of the vocal muscles which produces this staccato style, is analogous to the muscular action which produces the sharp, decisive, energetic movements of body indicating these states of mind; and therefore it is that the staccato style has the meaning we ascribe to it. Conversely, slurred intervals are expressive of gentler and less active feelings; and are so because they imply the smaller muscular vivacity due to a lower mental energy. The difference of effect resulting from difference of *time* in music, is also attributable to the same law. Already it has been pointed out that the more frequent changes of pitch which ordinarily result from passion, are imitated and developed in song; and here we have to add, that the various rates of such changes, appropriate to the different styles of music, are further traits having the same derivation. The slowest movements, *largo* and *adagio,* are used where such depressing emotions as grief, or such unexciting emotions as reverence, are to be portrayed; while the more rapid movements, *andante, allegro, presto,* represent successively increasing degrees of mental vivacity; and do this because they imply that muscular activity which flows from this mental vivacity. Even the *rhythm,* which forms a remaining distinction between song and speech, may not improbably have a kindred cause. Why the actions excited by strong feeling should tend to become rhythmical, is not very obvious; but that they do so there are divers evidences. There is the swaying of the body to and fro under pain or grief, of the leg under impatience or agitation. Dancing, too, is a rhythmical action natural to elevated emotion. That under excitement speech acquires a certain rhythm, we may occasionally perceive in the highest efforts of an orator. In poetry, which is a form of speech used for the better expression of emotional ideas, we have this rhythmical tendency developed. And when we bear in mind that dancing, poetry, and music are connate — are originally constituent parts of the same thing, it becomes clear that the measured movement common to them all implies a rhythmical action of the whole system, the vocal apparatus included; and that so the rhythm of music is a more subtle and complex result of this relation between mental and muscular excitement.

But it is time to end this analysis, which, possibly we have already carried too far. It is not to be supposed that the more special peculiarities of musical expression are to be definitely explained. Though probably they may all in some way conform to the principle that has been worked cut, it is obviously impracticable to trace that principle in its more ramified applications. Nor is it needful to our argument that it should be so traced. The foregoing facts sufficiently prove that what we regard as the distinctive traits of song, are simply the traits of emotional speech intensified and systematized. In respect of its general characteristics, we think it has been made clear that vocal music, and by consequence all music, is an idealization of the natural language of passion.

As far as it goes, the scanty evidence furnished by history confirms this conclusion. Note first the fact (not properly an historical one, but fitly grouped with such) that the dance-chants of savage tribes are very monotonous; and in virtue of their monotony are much more nearly allied to ordinary speech than are the songs of civilized races. Joining with this the fact that there are still extant among boatmen and others in the East, ancient chants of a like monotonous character, we may infer that vocal music originally diverged from emotional speech in a gradual, unobtrusive manner; and this is the inference to which our argument points. Further evidence to the same effect is supplied by Greek history. The early poems of the Greeks — which, be it remembered, were sacred legends embodied in that rhythmical, metaphorical language which strong feeling excites — were not recited, but chanted: the tones and the cadences were made musical by the same influences which made the speech poetical.

By those who have investigated the matter, this chanting is believed to have been not what we call singing, but nearly allied to our recitative; (far simpler indeed, if we may judge from the fact that the early Greek lyre, which had but *four* strings, was played in *unison* with the voice, which was therefore confined to four notes;) and as such, much less remote from common speech than our own singing is. For recitative, or musical recitation, is in all respects intermediate between speech and song. Its average effects are not so *loud* as those of song. Its tones are less sonorous in *timbre* than those of song. Commonly it diverges to a smaller extent from the middle notes — uses notes neither so high nor so low in *pitch.* The *intervals* habitual to it are neither so wide nor so varied. Its *rate of variation* is not so rapid. And at the same time that its primary *rhythm* is less decided, it has none of that secondary rhythm produced by recurrence of the same or parallel musical phrases, which is one of the marked characteristics of song. Thus, then, we may not only infer, from the evidence furnished by existing barbarous tribes, that the vocal music of pre-historic times was emotional speech very slightly exalted; but we see that the earliest vocal music of which we have any account, differed much less from emotional speech than does the vocal music of our days.

That recitative — beyond which, by the way, the Chinese and Hindoos seem never to have advanced — grew naturally out of the modulations and cadences of strong feeling, we have indeed still current evidence. There are even now to be met with occasions on which strong feeling vents itself in this form. Whoever has been present when a meeting of Quakers was addressed by one of their preachers (whose practice it is to speak only under the influence of religious emotion), must have been struck by the quite unusual tones, like those of a subdued chant, in which the address was made. It is clear, too, that the intoning used in some churches, is representative of this same mental state; and has been adopted on account of the instinctively felt congruity between it and the contrition, supplication, or reverence verbally expressed.

And if, as we have good reason to believe, recitative arose by degrees out of emotional speech, it becomes manifest that by a continuance of the same process song has arisen out of recitative. Just as, from the orations and legends of savages, expressed in the metaphorical, allegorical style natural to them, there sprung epic poetry, out of which lyric poetry was afterwards developed; so, from the exalted tones and cadences in which such orations and legends were delivered, came the chant or recitative music, from whence lyrical music has since grown up. And there has not only thus been a simultaneous and parallel genesis, but there is also a parallelism of results. For lyrical poetry differs from epic poetry, just as lyrical music differs from recitative: each still further intensifies the natural language of the emotions.

Lyrical poetry is more metaphorical, more hyperbolic, more elliptical, and adds the rhythm of lines to the rhythm of feet; just as lyrical music is louder, more sonorous, more extreme in its intervals, and adds the rhythm of phrases to the rhythm of bars. And the known fact that out of epic poetry the stronger passions developed lyrical poetry as their appropriate vehicle, strengthens the inference that they similarly developed lyrical music out of recitative.

Nor indeed are we without evidences of the transition. It needs but to listen to an opera to hear the leading gradations. Between the comparatively level recitative of ordinary dialogue, the more varied recitative with wider intervals and higher tones used in exciting scenes, the still more musical recitative which preludes an air, and the air itself, the successive steps are but small; and the fact that among airs themselves gradations of like nature may be traced, further confirms the conclusion that the highest form of vocal music was arrived at by degrees.

Moreover, we have some clue to the influences which have induced this development; and may roughly conceive the process of it. As the tones, intervals, and cadences of strong emotion were the elements out of which song was elaborated; so, we may expect to find that still stronger emotion produced the elaboration: and we have evidence implying this. Instances in abundance may be cited, showing that musical composers are men of extremely acute sensibilities. The Life of Mozart depicts him as one of intensely active affections and highly impressionable temperament. Various anecdotes represent Beethoven as very susceptible and very passionate. Mendelssohn is described by those who knew him to have been full of fine feeling. And the almost incredible sensitiveness of Chopin has been illustrated in the memoirs of George Sand. An unusually emotional nature being thus the general characteristic of musical composers, we have in it just the agency required for the development of recitative and song. Intenser feeling producing intenser manifestations, any cause of excitement will call forth from such a nature, tones and changes of voice more marked than those called forth from an ordinary nature — will generate just those exaggerations which we have found to distinguish the lower vocal music from emotional speech, and the higher vocal music from the lower. Thus it becomes credible that the four-toned recitative of the early Greek poets (like all poets, nearly allied to composers in the comparative intensity of their feelings), was really nothing more than the slightly exaggerated emotional speech natural to them, which grew by frequent use into an organized form. And it is readily conceivable that the accumulated agency of subsequent poet-musicians, inheriting and adding to the products of those who went before them, sufficed, in the course of the ten centuries which we know it took, to develope this four-toned recitative into a vocal music having a range of two octaves.

Not only may we so understand how more sonorous tones, greater extremes of pitch, and wider intervals, were gradually introduced; but also how there arose a greater variety and complexity of musical expression. For this same passionate, enthusiastic temperament, which naturally leads the musical composer to express the feelings possessed by others as well as himself, in extremer intervals and more marked cadences than they would use, also leads him to give musical utterance to feelings which they either do not experience, or experience in but slight degrees. In virtue of this general susceptibility which distinguishes him, he regards with emotion, events, scenes, conduct, character, which produce upon most men no appreciable effect. The emotions so generated, compounded as they are of the simpler emotions, are not expressible by intervals and cadences natural to these, but by combinations of such intervals and cadences: whence arise more involved musical phrases, conveying more complex, subtle, and unusual feelings. And thus we may in some measure understand how it happens that music not only so strongly excites our more familiar feelings, but also produces feelings we never had before — arouses dormant sentiments of which we had not conceived the possibility and do not know the meaning; or, as Richter says — tells us of things we have not seen and shall not see.

Indirect evidences of several kinds remain to be briefly pointed out. One of them is the difficulty, not to say impossibility, of otherwise accounting for the expressiveness of music. Whence comes it that special combinations of notes should have special effects upon our emotions? — that one should give us a feeling of exhilaration, another of melancholy, another of affection, another of reverence? Is it that these special combinations have intrinsic meanings apart from the human constitution? — that a certain number of aerial waves per second, followed by a certain other number, in the nature of things signify grief, while in the reverse order they signify joy; and similarly with all other intervals, phrases, and cadences? Few will be so irrational as to think this. Is it, then, that the meanings of these special combinations are conventional only? — that we learn their implications, as we do those of words, by observing how others understand them? This is an hypothesis not only devoid of evidence, but directly opposed to the experience of every one. How, then, are musical effects to be explained? If the theory above set forth be accepted, the difficulty disappears. If music, taking for its raw material the various modifications of voice which are the physiological results of excited feeling, intensifies, combines, and complicates them — if it exaggerates the loudness, the resonance, the pitch, the intervals, and the variability, which, in virtue of an organic law, are the characteristics of passionate speech — if, by carrying out these further, more consistently, more unitedly, and more sustainedly, it produces an idealized language of emotion; then its power over us becomes comprehensible. But in the absence of this theory, the expressiveness of music appears to be inexplicable.

Again, the preference we feel for certain qualities of sound presents a like difficulty, admitting only of a like solution. It is generally agreed that the tones of the human voice are more pleasing than any others. Grant that music takes its rise from the modulations of the human voice under emotion, and it becomes a natural consequence that the tones of that voice should appeal to our feelings more than any others; and so should be considered more beautiful than any others. But deny that music has this origin, and the only alternative is the untenable position that the vibrations proceeding from a vocalist's throat are, objectively considered, of a higher order than those from a horn or a violin. Similarly with harsh and soft sounds. If the conclusiveness of the foregoing reasonings be not admitted, it must be supposed that the vibrations causing the last are intrinsically better than those causing the first; and that, in virtue of some pre-established harmony, the higher feelings and natures produce the one, and the lower the other. But if the foregoing reasonings be valid, it follows, as a matter of course, that we shall like the sounds that habitually accompany agreeable feelings, and dislike those that habitually accompany disagreeable feelings.

Once more, the question — How is the expressiveness of music to be otherwise accounted for? may be supplemented by the question — How is the genesis of music to be otherwise accounted for? That music is a product of civilization is manifest; for though savages have their dance-chants, these are of a kind scarcely to be dignified by the title musical: at most, they supply but the vaguest rudiment of music, properly so called. And if music has been by slow steps developed in the course of civilization, it must have been developed out of something. If, then, its origin is not that above alleged, what is its origin?

Thus we find that the negative evidence confirms the positive, and that, taken together, they furnish strong proof. We have seen that there is a physiological relation, common to man and all animals, between feeling and muscular action; that as vocal sounds are produced by muscular action, there is a consequent physiological relation between feeling and vocal sounds; that all the modifications of voice expressive of feeling are the direct results of this physiological relation; that music, adopting all these modifications, intensifies them more and more as it ascends to its higher and higher forms, and becomes music simply in virtue of thus intensifying them; that, from the ancient epic poet chanting his verses, down to the modern musical composer, men of unusually strong feelings prone to express them in extreme forms, have been naturally the agents of these successive intensifications; and that so there has little by little arisen a wide divergence between this idealized language of emotion and its natural language: to which direct evidence we have just added the indirect — that on no other tenable hypothesis can either the expressiveness or the genesis of music be explained.

And now, what is the *function* of music? Has music any effect beyond the immediate pleasure it produces? Analogy suggests that it has. The enjoyments of a good dinner do not end with themselves, but minister to bodily well-being. Though people do not marry with a view to maintain the race, yet the passions which impel them to marry secure its maintenance. Parental affection is a feeling which, while it conduces to parental happiness, ensures the nurture of offspring. Men love to accumulate property, often without thought of the benefits it produces; but in pursuing the pleasure of acquisition they indirectly open the way to other pleasures. The wish for public approval impels all of us to do many things which we should otherwise not do, — to undertake great labours, face great dangers, and habitually rule ourselves in a way that smooths social intercourse: that is, in gratifying our love of approbation we subserve divers ulterior purposes. And, generally, our nature is such that in fulfilling each desire, we in some way facilitate the fulfilment of the rest. But the love of music seems to exist for its own sake. The delights of melody and harmony do not obviously minister to the welfare either of the individual or of society. May we not suspect, however, that this exception is apparent only? Is it not a rational inquiry — What are the indirect benefits which accrue from music, in addition to the direct pleasure it gives?

But that it would take us too far out of our track, we should prelude this inquiry by illustrating at some length a certain general law of progress; — the law that alike in occupations, sciences, arts, the divisions that had a common root, but by continual divergence have become distinct, and are now being separately developed, are not truly independent, but severally act and react on each other to their mutual advancement. Merely hinting thus much, however, by way of showing that there are many analogies to justify us, we go on to express the opinion that there exists a relationship of this kind between music and speech.

All speech is compounded of two elements, the words and the tones in which they are uttered — the signs of ideas and the signs of feelings. While certain articulations express the thought, certain vocal sounds express the more or less of pain or pleasure which the thought gives. Using the word *cadence* in an unusually extended sense, as comprehending all modifications of voice, we may say that *cadence is the commentary of the emotions upon the propositions of the intellect.* This duality of spoken language, though not formally recognised, is recognised in practice by every one; and every one knows that very often more weight attaches to the tones than to the words. Daily experience supplies cases in which the same sentence of disapproval will be understood as meaning little or meaning much, according to the inflections of voice which accompany it; and daily experience supplies still more striking cases in which words and tones are in direct contradiction — the first expressing consent, while the last express reluctance; and the last being believed rather than the first.

These two distinct but interwoven elements of speech have been undergoing a simultaneous development. We know that in the course of civilization words have been multiplied, new parts of speech have been introduced, sentences have grown more varied and complex; and we may fairly infer that during the same time new modifications of voice have come into use, fresh intervals have been adopted, and cadences have become more elaborate. For while, on the one hand, it is absurd to suppose that, along with the undeveloped verbal forms of barbarism, there existed a developed system of vocal inflections; it is, on the other hand, necessary to suppose that, along with the higher and more numerous verbal forms needed to convey the multiplied and complicated ideas of civilized life, there have grown up those more involved changes of voice which express the feelings proper to such ideas. If intellectual language is a growth, so also, without doubt, is emotional language a growth.

Now, the hypothesis which we have hinted above, is, that beyond the direct pleasure which it gives, music has the indirect effect of developing this language of the emotions. Having its root, as we have endeavoured to show, in those tones, intervals, and cadences of speech which express feeling — arising by the combination and intensifying of these, and coming finally to have an embodiment of its own; music has all along been reacting upon speech, and increasing its power of rendering emotion. The use in recitative and song of inflections more expressive than ordinary ones, must from the beginning have tended to develope the ordinary ones. Familiarity with the more varied combinations of tones that occur in vocal music, can scarcely have failed to give greater variety of combination to the tones in which we utter our impressions and desires. The complex musical phrases by which composers have conveyed complex emotions, may rationally be supposed to have influenced us in making those involved cadences of conversation by which we convey our subtler thoughts and feelings.

That the cultivation of music has no effect on the mind, few will be absurd enough to contend. And if it has an effect, what more natural effect is there than this of developing our perception of the meanings of inflections, qualities, and modulations of voice; and giving us a correspondingly increased power of using them? Just as mathematics, taking its start from the phenomena of physics and astronomy, and presently coming to be a separate science, has since reacted on physics and astronomy to their immense advancement — just as chemistry, first arising out of the processes of metallurgy and the industrial arts, and gradually growing into an independent study, has now become an aid to all kinds of production — just as physiology, originating out of medicine and once subordinate to it, but latterly pursued for its own sake, is in our day coming to be the science on which the progress of medicine depends; — so, music, having its root in emotional language, and gradually evolved from it, has ever been reacting upon and further advancing it. Whoever will examine the facts, will find this hypothesis to be in harmony with the method of civilization everywhere displayed.

It will scarcely be expected that much direct evidence in support of this conclusion can be given. The facts are of a kind which it is difficult to measure, and of which we have no records. Some suggestive traits, however, may be noted. May we not say, for instance, that the Italians, among whom modern music was earliest cultivated, and who have more especially practised and excelled in melody (the division of music with which our argument is chiefly concerned) — may we not say that these Italians speak in more varied and expressive inflections and cadences than any other nation? On the other hand, may we not say that, confined almost exclusively as they have hitherto been to their national airs, which have a marked family likeness, and therefore accustomed to but a limited range of musical expression, the Scotch are unusually monotonous in the intervals and modulations of their speech? And again, do we not find among different classes of the same nation, differences that have like implications? The gentleman and the clown stand in very decided contrast with respect to variety of intonation. Listen to the conversation of a servant-girl, and then to that of a refined, accomplished lady, and the more delicate and complex changes of voice used by the latter will be conspicuous. Now, without going so far as to say that out of all the differences of culture to which the upper and lower classes are subjected, difference of musical culture is that to which alone this difference of speech is ascribable; yet we may fairly say that there seems a much more obvious connexion of cause and effect between these than between any others. Thus, while the inductive evidence to which we can appeal is but scanty and vague, yet what there is favours our position.

Probably most will think that the function here assigned to music is ore of very little moment. But further reflection may load them to a contrary conviction. In its bearings upon human happiness, we believe that this emotional language which musical culture developes and refines, is only second in importance to the language of the intellect; perhaps not even second to it. For these modifications of voice produced by feelings, are the means of exciting like feelings in others. Joined with gestures and expressions of face, they give life to the otherwise dead words in which the intellect utters its ideas; and so enable the hearer not only to *understand* the state of mind they accompany, but to *partake* of that state. In short, they are the chief media of *sympathy.* And if we consider how much both our general welfare and our immediate pleasures depend upon sympathy, we shall recognise the importance of whatever makes this sympathy greater. If we bear in mind that by their fellow-feeling men are led to be have justly, kindly and considerately to each other — that the difference between the cruelty of the barbarous and the humanity of the civilized, results from the increase of fellow-feeling; if we bear in mind that this faculty which makes us sharers in the joys and sorrows of others, is the basis of all the higher affections — that in friendship, love, and all domestic pleasures, it is an essential element; if we bear in mind how much our direct gratifications are intensified by sympathy, — how, at the theatre, the concert, the picture gallery, we lose half our enjoyment if we have no one to enjoy with us; if, in short, we bear in mind that for all happiness beyond what the unfriended recluse can have, we are indebted to this same sympathy; — we shall see that the agencies which communicate it can scarcely be overrated in value.

The tendency of civilization is more and more to repress the antagonistic elements of our characters and to develope the social ones — to curb our purely selfish desires and exercise our unselfish ones — to replace private gratifications by gratifications resulting from, or involving, the happiness of others. And while, by this adaptation to the social state, the sympathetic side of our nature is being unfolded, there is simultaneously growing up a language of sympathetic intercourse — a language through which we communicate to others the happiness we feel, and are made sharers in their happiness.

This double process, of which the effects are already sufficiently appreciable, must go on to an extent of which we can as yet have no adequate conception. The habitual concealment of our feelings diminishing, as it must, in proportion as our feelings become such as do not demand concealment, we may conclude that the exhibition of them will become much more vivid than we now dare allow it to be; and this implies a more expressive emotional language. At the same time, feelings of a higher and more complex kind, as yet experienced only by the cultivated few, will become general; and there will be a corresponding development of the emotional language into more involved forms. Just as there has silently grown up a language of ideas, which, rude as it at first was, now enables us to convey with precision the most subtle and complicated thoughts; so, there is still silently growing up a language of feelings, which notwithstanding its present imperfection, we may expect will ultimately enable men vividly and completely to impress on each other all the emotions which they experience from moment to moment.

Thus if, as we have endeavoured to show, it is the function of music to facilitate the development of this emotional language, we may regard music as an aid to the achievement of that higher happiness which it indistinctly shadows forth. Those vague feelings of unexperienced felicity which music arouses — those indefinite impressions of an unknown ideal life which it calls up, may be considered as a prophecy, to the fulfilment of which music is itself partly instrumental. The strange capacity which we have for being so affected by melody and harmony, may be taken to imply both that it is within the possibilities of our nature to realize those intenser delights they dimly suggest, and that they are in some way concerned in the realization of them. On this supposition the power and the meaning of music become comprehensible; but otherwise they are a mystery.

We will only add, that if the probability of these corollaries be admitted, then music must take rank as the highest of the fine arts — as the one which, more than any other, ministers to human welfare. And thus, even leaving out of view the immediate gratifications it is hourly giving, we cannot too much applaud that progress of musical culture which is becoming one of the characteristics of our age.

VI
THE NEBULAR HYPOTHESIS

Inquiring into the pedigree of an idea is not a bad means of roughly estimating its value. To have come of respectable ancestry, is *primâ facie* evidence of worth in a belief as in a person; while to be descended from a discreditable stock is, in the one case as in the other, an unfavorable index. The analogy is not a mere fancy. Beliefs, together with those who hold them, are modified little by little in successive generations; and as the modifications which successive generations of the holders undergo, do not destroy the original type, but only disguise and refine it, so the accompanying alterations of belief, however much they purify, leave behind the essence of the original belief.

Considered genealogically, the received theory respecting the creation of the Solar System is unmistakeably of low origin. You may clearly trace it back to primitive mythologies. Its remotest ancestor is the doctrine that the celestial bodies are personages who originally lived on the Earth — a doctrine still held by some of the negroes Livingstone visited. Science having divested the sun and planets of their divine personalities, this old idea was succeeded by the idea which even Kepler entertained, that the planets are guided in their courses by presiding spirits: no longer themselves gods, they are still severally kept in their orbits by gods. And when gravitation came to dispense with these celestial steersmen, there was begotten a belief, less gross than its parent, but partaking of the same essential nature, that the planets were originally launched into their orbits from the Creator's hand. Evidently, though much refined, the anthropomorphism of the current hypothesis is inherited from the aboriginal anthropomorphism, which described gods as a stronger order of men.

There is an antagonist hypothesis which does not propose to honour the Unknown Power manifested in the Universe, by such titles as "The Master-Builder," or "The Great Artificer;" but which regards this Unknown Power as probably working after a method quite different from that of human mechanics. And the genealogy of this hypothesis is as high as that of the other is low. It is begotten by that ever-enlarging and ever-strengthening belief in the presence of Law, which accumulated experiences have gradually produced in the human mind. From generation to generation Science has been proving uniformities of relation among phenomena which were before thought either fortuitous or supernatural in their origin — has been showing an established order and a constant causation where ignorance had assumed irregularity and arbitrariness. Each further discovery of Law has increased the presumption that Law is everywhere conformed to. And hence, among other beliefs, has arisen the belief that the Solar System originated, not by *manufacture* but by *evolution.* Besides its abstract parentage in those grand general conceptions which positive Science has generated, this hypothesis has a concrete parentage of the highest character. Based as it is on the law of universal gravitation, it may claim for its remote progenitor the great thinker who established that law. The man who gave it its general shape, by promulgating the doctrine that stars result from the aggregation of diffused matter, was the most diligent, careful, and original astronomical observer of modern times. And the world has not seen a more learned mathematician than the man who, setting out with this conception of diffused matter concentrating towards its centre of gravity, pointed out the way in which there would arise, in the course of its concentration, a balanced group of sun, planets, and satellites, like that of which the Earth is a member.

Thus, even were there but little direct evidence assignable for the Nebular Hypothesis, the probability of its truth would still be strong. Its own high derivation and the low derivation of the antagonist hypothesis, would together form a weighty reason for accepting it — at any rate, provisionally. But the direct evidence assignable for the Nebular Hypothesis is by no means little. It is far greater in quantity, and more varied in kind, than is commonly supposed. Much has been said here and there on this or that class of evidences; but nowhere, as far as we know, have all the evidences, even of one class, been fully stated; and still less has there been an adequate statement of the several groups of evidences in their *ensemble.* We propose here to do something towards supplying the deficiency: believing that, joined with the *à priori* reasons given above, the array of *à posteriori* reasons will leave little doubt in the mind of any candid inquirer.

And first, let us address ourselves to those recent discoveries in stellar astronomy, which have been supposed to conflict with this celebrated speculation.

When Sir William Herschel, directing his great reflector to various nebulous spots, found them resolvable into clusters of stars, he inferred, and for a time maintained, that all nebulous spots are clusters of stars exceedingly remote from us. But after years of conscientious investigation, he concluded that "there were nebulosities which are not of a starry nature;" and on this conclusion was based his hypothesis of a diffused luminous fluid, which by its eventual aggregation, produced stars. A telescopic power much exceeding that used by Herschel, has enabled Lord Rosse to resolve some of the nebulæ previously unresolved; and, returning to the conclusion which Herschel first formed on similar grounds but afterwards rejected, many astronomers have assumed that, under sufficiently high powers, every nebula would be decomposed into stars — that the resolvability is solely a question of distance. The hypothesis now commonly entertained is, that all nebulæ are galaxies more or less like in nature to that immediately surrounding us; but that they are so inconceivably remote, as to look, through an ordinary telescope, like small faint spots. And not a few have drawn the corollary, that by the discoveries of Lord Rosse the Nebular Hypothesis has been disproved.

Now, even supposing that these inferences respecting the distances and natures of the nebulæ are valid, they leave the Nebular Hypothesis substantially as it was. Admitting that each of those faint spots is a sidereal system, so far removed that its countless stars give less light than one small star of our own sidereal system; the admission is in no way inconsistent with the belief, that stars and their attendant planets have been formed by the aggregation of nebulous matter. Though, doubtless, if the existence of nebulous matter now in course of concentration be disproved, one of the evidences of the Nebular Hypothesis is destroyed; yet the remaining evidences remain just as they were. It is a perfectly tenable position, that though nebular condensation is now nowhere to be seen in progress, yet it was once going on universally. And, indeed, it might be argued that the still-continued existence of diffused nebulous matter is scarcely to be expected; seeing that the causes which have resulted in the aggregation of one mass, must have been acting on all masses, and that hence the existence of masses not aggregated would be a fact calling for explanation. Thus, granting the immediate conclusions suggested by these recent disclosures of the six-feet reflector, the corollary which many have drawn is inadmissible.

But we do not grant these conclusions. Receiving them though we have, for years past, as established truths, a critical examination of the facts has convinced us that they are quite unwarrantable. They involve so many manifest incongruities, that we have been astonished to find men of science entertaining them even as probable hypotheses. Let us consider these incongruities.

In the first place, mark what is inferable from the distribution of nebulæ.

"The spaces which precede or which follow simple nebulæ," says Arago, "and, *à fortiori,* groups of nebulæ, contain generally few stars. Herschel found this rule to be invariable. Thus, every time that, during a short interval, no star approached, in virtue of the diurnal motion, to place itself in the field of his motionless telescope, he was accustomed to say to the secretary who assisted him, 'Prepare to write; nebulæ are about to arrive.' "

How does this fact consist with the hypothesis that nebulæ are remote galaxies? If there were but one nebula, it would be a curious coincidence were this one nebula so placed in the distant regions of space, as to agree in direction with a starless spot in our own sidereal system. If there were but two nebulæ, and both were so placed, the coincidence would be excessively strange. What, then, shall we say on finding that there are thousands of nebulæ so placed? Shall we believe that in thousands of cases these far-removed galaxies happen to agree in their visible positions with the thin places in our own galaxy? Such a belief is next to impossible. Still more manifest does the impossibility of it become when we consider the general distribution of nebulæ. Besides again showing itself in the fact that "the poorest regions in stars are near the richest in nebulæ," the law above specified applies to the heavens as a whole. In that zone of celestial space where stars are excessively abundant, nebulæ are rare; while in the two opposite celestial spaces that are furthest removed from this tone, nebulæ are abundant. Scarcely any nebulæ lie near the galactic circle (or plane of the Milky Way); and the great mass of them lie round the galactic poles. Can this also be mere coincidence? When to the fact that the general mass of nebulæ are antithetical in position to the general mass of stars, we add the fact that local regions of nebulæ are regions where stars are scarce, and the further fact that single nebulæ are habitually found in comparatively starless spots; does not the proof of a physical connexion become overwhelming? Should it not require an infinity of evidence to show that nebulas are not parts of our sidereal system? Let us see whether any such infinity of evidence is assignable. Let us see whether there is even a single alleged proof which will bear examination.

"As seen through colossal telescopes," says Humboldt, "the contemplation of these nebulous masses leads us into regions from whence a ray of light, according to an assumption not wholly improbable, requires millions of years to reach our earth — to distances for whose measurement the dimensions (the distance of Sirius, or the calculated distances of the binary stars in Cygnus and the Centaur) of our nearest stratum of fixed stars scarcely suffice."

Now, in this somewhat confused sentence there is expressed a more or less decided belief, that the distances of the nebulæ from our galaxy of stars as much transcend the distances of our stars from each other, as these interstellar distances transcend the dimensions of our planetary system. Just as the diameter of the Earth's orbit, is an inappreciable point when compared with the distance of our Sun from Sirius; so is the distance of our Sun from Sirius, an inappreciable point when compared with the distance of our galaxy from those far removed galaxies constituting nebulæ. Observe the consequences of this assumption.

If one of these supposed galaxies is so remote that its distance dwarfs our interstellar spaces into points, and therefore makes the dimensions of our whole sidereal system relatively insignificant; does it not inevitably follow that the telescopic power required to resolve this remote galaxy into stars, must be incomparably greater than the telescopic power required to resolve the whole of our own galaxy into stars? Is it not certain that an instrument which can just exhibit with clearness the most distant stars of our own cluster, must be utterly unable to separate one of these remote clusters into stars? What, then, are we to think when we find that the same instrument which decomposes hosts of nebulæ into stars, *fails* to resolve completely our own Milky Way? Take a homely comparison. Suppose a man surrounded by a swarm of bees, extending, as they sometimes do, so high in the air as to be individually almost invisible, were to declare that a certain spot on the horizon was a swarm of bees; and that he knew it because he could see the bees as separate specks. Astounding as the assertion would be, it would not exceed in incredibility this which we are criticising. Reduce the dimensions to figures, and the absurdity becomes still more palpable. In round numbers, the distance of Sirius from the Earth is a million times the distance of the Earth from the Sun; and, according to the hypothesis, the distance of a nebula is something like a million times the distance of Sirius.

Now, our own "starry island, or nebula," as Humboldt calls it, "forms a lens-shaped, flattened, and everywhere detached stratum, whose major axis is estimated at seven or eight hundred, and its minor axis at a hundred and fifty times the distance of Sirius from the Earth."[9] And since it is concluded that our Solar System is near the centre of this aggregation, it follows that our distance from the remotest parts of it is about four hundred distances of Sirius. But the stars forming these remotest parts are not individually visible, even through telescopes of the highest power. How, then, can such telescopes make individually visible the stars of a nebula which is a million times the distance of Sirius? The implication is, that a star rendered invisible by distance becomes visible if taken two thousand five hundred times further off! Shall we accept this implication? or shall we not rather conclude that the nebulæ are *not* remote galaxies? Shall we not infer that, be their nature what it may, they must be at least as near to us as the extremities of our own sidereal system?

Throughout the above argument, it is tacitly assumed that differences of apparent magnitude among the stars, result mainly from differences of distance. On this assumption the current doctrines respecting the nebulæ are founded; and this assumption is, for the nonce, admitted in each of the foregoing criticisms. From the time, however, when it was first made by

[9] Cosmos. (Seventh Edition.) Vol. i. pp. 79, 80.

Sir W. Herschel, this assumption has been purely gratuitous; and it now proves to be totally inadmissible. But, awkwardly enough, its truth and its untruth are alike fatal to the conclusions of those who argue after the manner of Humboldt. Note the alternative.

On the one hand, what follows from the untruth of the assumption? If apparent largeness of stars is not due to comparative nearness, and their successively smaller sizes to their greater and greater degrees of remoteness, what becomes of the inferences respecting the dimensions of our sidereal system and the distances of nebulæ? If, as has lately been shown, the almost invisible star 61 Cygni has a greater parallax than a Cygni, though, according to an estimate based on Sir W. Herschel's assumption, it should be about twelve times more distant — if, as it turns out, there exist telescopic stars which are nearer to us than Sirius; of what worth is the conclusion that the nebulæ are very re mote, because their component luminous masses are made visible only by high telescopic powers? Clearly, if the most brilliant star in the heavens and a star that cannot be seen by the naked eye, prove to be equidistant, relative distances cannot be in the least inferred from relative visibilities. And if so, nebulæ may be comparatively near, though the starlets of which they are made up appear extremely minute.

On the other hand, what follows if the truth of the assumption be granted? The arguments used to justify this assumption in the case of the stars, equally justify it in the case of the nebulæ. It cannot be contended that, on the average, the *apparent* sizes of the stars indicate their distances, without its being admitted that, on the average, the *apparent* sizes of the nebulæ indicate their distances — that, generally speaking, the larger are the nearer, and the smaller are the more distant. Mark, now, the necessary inference respecting their resolvability. The largest or nearest nebulæ will be most easily resolved into stars; the successively smaller will be successively more difficult of resolution; and the irresolvable ones will be the smallest ones. This, however, is exactly the reverse of the fact. The largest nebulæ are either wholly irresolvable, or but partially resolvable under the highest telescopic powers; while a great proportion of quite small nebulæ, are easily resolved by far less powerful telescopes. An instrument through which the great nebula in Andromeda, two and a half degrees long and one degree broad, appears merely as a diffused light, decomposes a nebula of fifteen minutes diameter into twenty thousand starry points. At the same time that the individual stars of a nebula eight minutes in diameter are so clearly seen as to allow of their number being estimated, a nebula covering an area five hundred times as great shows no stars at all. What possible explanation can be given of this on the current hypothesis?

Yet a further difficulty remains — one which is, perhaps, still more obviously fatal than the foregoing. This difficulty is presented by the phenomena of the Magellanic clouds. Describing the larger of these, Sir John Herschel says: —

"The nubecula major, like the minor, consists partly of large tracts and ill-defined patches of irresolvable nebula, and of nebulosity in every stage of resolution, up to perfectly resolved stars like the Milky Way; as also of regular and irregular nebula properly so called, of globular clusters in every stage of resolvability, and of clustering groups sufficiently insulated and condensed to come under the designation of 'cluster of stars.' " — "Cape Observations," p.146.

In his "Outlines of Astronomy," Sir John Herschel, after repeating this description in other words, goes on to remark that —

"This combination of characters, rightly considered, is in a high degree instructive, affording an insight into the probable comparative distance of *stars* and *nebulæ,* and the real brightness of individual stars as compared with one another. Taking the apparent semi-diameter of the nubecula major at three degrees, and regarding its solid form as, roughly speaking, spherical, its nearest and most remote parts differ in their distance from us by a little more than a tenth part of our distance from its centre. The brightness of objects situated in its nearer portions, therefore, cannot be *much* exaggerated, nor that of its remoter *much* enfeebled, by their difference of distance. Yet within this globular space we have collected upwards of six hundred stars of the seventh, eighth, ninth, and tenth magnitude, nearly three hundred nebulæ, and globular and other clusters *of all degrees of resolvability,* and smaller scattered stars of every inferior magnitude, from the tenth to such as by their magnitude and minuteness constitute irresolvable nebulosity, extending over tracts of many square degrees. Were there but one such object, it might be maintained without utter improbability that its apparent sphericity is only an effect of foreshortening, and that in reality a much greater proportional difference of distance between its nearer and more remote parts exists. But such an adjustment, improbable enough in one case, must be rejected as too much so for fair argument in two. It must, therefore, be taken as a demonstrated fact, that stars of the seventh or eighth magnitude, and irresolvable nebula, may co-exist within limits of distance not differing in proportion more than as nine to ten." — "Outlines of Astronomy," pp. 614, 615.

Now, we think this supplies a *reductio ad absurdum* of the doctrine we are combating. It gives us the choice of two incredibilities. If we are to believe that one of these nebulæ is so remote that its hundred thousand stars look like a milky spot, invisible to the naked eye; we must also believe that there are single stars so enormous that though removed to this same distance they remain visible. If we accept the other alternative, and say that many nebulæ are no further off than our own stars of the eighth magnitude; then it is requisite to say that at a distance not greater than that at which a single star is still faintly visible to the naked eye, there may exist a group of a hundred thousand stars which is invisible to the naked eye. Neither of these positions can be entertained. What, then, is the conclusion that remains? This, only: — that the nebula are not further off from us than parts of our own sidereal system, of which they must be considered members; and that when they are resolvable into discrete masses, these masses cannot be considered as stars in anything like the ordinary sense of that word.

And now, having seen the untenability of this idea, rashly espoused by sundry astronomers, that the nebulæ are extremely remote galaxies; let us consider whether the various appearances they present are not reconcile able with the Nebular Hypothesis.

Given a rare and widely-diffused mass of nebulous matter, having a diameter, say as great as the distance from the Sun to Sirius,[10] what are the successive changes that will take place in it? Mutual gravitation will approximate its atoms; but their approximation will be opposed by atomic repulsion, the overcom-

---

10 Any objection made to the extreme tenuity this involves, is met by the calculation of Newton, who proved that were a spherical rich of air removed four thousand miles from the Earth, it would expend into a sphere more than filling the orbit of Saturn.

ing of which implies the evolution of heat. As fast as this heat partially escapes by radiation, further approximation will take place, attended by further evolution of heat, and so on continuously: the processes not occurring separately as here described, but simultaneously, uninterruptedly, and with increasing activity. Eventually, this slow movement of the atoms towards their common centre of gravity, will bring about phenomena of another order.

Arguing from the known laws of atomic combination, it will happen that when the nebulous mass has reached a particular stage of condensation — when its internally-situated atoms have approached to within certain distances, have generated a certain amount of heat, and are subject to a certain mutual pressure (the heat and pressure both increasing as the aggregation progresses); some of them will suddenly enter into chemical union. Whether the binary atoms so produced be of kinds such as we know, which is possible; or whether they be of kinds simpler thin any we know, which is more probable; matters not to the argument. It suffices that molecular combination of some species will finally take place. When it does take place, it will be accompanied by a great and sudden disengagement of heat; and until this excess of heat has escaped, the newly-formed binary atoms will remain uniformly diffused, or, as it were, dissolved in the preexisting nebulous medium.

But now mark what must by-and-by happen. When radiation has adequately lowered the temperature, these binary atoms will precipitate; and having precipitated, they will not remain uniformly diffused, but will aggregate into *flocculi:* just as water, precipitated from air, collects into clouds. This *à priori* conclusion is confirmed by the observation of those still extant portions of nebulous matter which constitute comets; for, "that the luminous part of a comet is something in the nature of a smoke, fog, or cloud, suspended in a transparent atmosphere, is evident," says Sir John Herschel.

Concluding, then, that a nebulous mass will, in course of time, resolve itself into flocculi of precipitated denser matter, floating in the rarer medium from which they were precipitated, let us inquire what will be the mechanical results. We shall find that they will be quite different from those occurring in the original homogeneous mass; and also quite different from those which would occur among discrete masses dispersed through empty space. Of clustered bodies in empty space, each will move along a line which is the resultant of the tractive forces exercised by all the rest, modified from moment to moment by the acquired motion. So, too, with bodies dispersed through a resisting medium, provided they are spherical. But irregular bodies dispersed through a resisting medium will not severally do this. A mass which presents an irregular face to its line of movement through a resisting medium, must be deflected from its otherwise-determined course, by the unequal reactions on its different sides. Hence the flocculi, as, by analogy, we term these precipitated masses of gas or vapour, will severally acquire movements not such as will make them meet directly in the common centre of gravity, but such as will carry them severally to one or other side of it; and their movements, accelerated as well as changed in direction by the increasing centripetal force, but retarded by the resisting medium, will result in something like spirals finally ending in the common centre of gravity. Observe, however, that this conclusion by no means implies a common spiral movement of all the flocculi; for as they must not only be varied in their forms, but disposed in all varieties of position, their respective movements will be deflected, not towards one side of the common centre of gravity, but towards various sides. How then can there result a spiral movement common to them all? Very simply. Each flocculus, in describing its spiral course, must give motion to the rarer medium through which it is moving.

Now, the probabilities are infinity to one against all the respective motions thus impressed on this rarer medium, exactly balancing each other. And if they do not balance each other, the inevitable result must be a rotation of the whole mass of the rarer medium in one direction. But preponderating momentum in one direction, having caused rotation of the medium in that direction, the rotating medium must in its turn gradually arrest such flocculi as are moving in opposition, and impress its own motion upon them; and thus there will ultimately be formed a rotating medium with suspended flocculi partaking of its motion, while they move in converging spirals towards the common centre of gravity.

Before comparing these conclusions with the facts, let us pursue the reasoning a little further, and observe the subordinate actions, and the endless modifications which will result from them. The respective flocculi must not only be drawn towards their common centre of gravity, but also towards neighbouring flocculi. Hence the whole assemblage of flocculi will break up into subordinate groups: each group concentrating towards its local centre of gravity, and in so doing acquiring a vertical movement, like that subsequently acquired by the whole nebula. Now, according to circumstances, and chiefly according to the size of the original nebulous mass, this process of local aggregation will produce various results. If the whole nebula is but small, the local groups of flocculi may be drawn into the common centre of gravity before their constituent masses have coalesced with each other. In a larger nebula, these local aggregations may have concentrated into rotating spheroids of vapour, while yet they have made but little approach towards the general focus of the system. In a still larger nebula, where the local aggregations are both greater and more remote from the common centre of gravity, they may have condensed into masses of molten matter before the general distribution of them has greatly altered. In short, as the conditions in each case determine, the discrete masses produced may vary indefinitely in number, in size, in density, in motion, in distribution.

And now let us return to the visible characters of the nebulæ, as observed through modern telescopes. Take first the description of those nebulæ which, by the hypothesis, must be in an early stage of evolution.

"Among the *irregular nebulœ*" says Sir John Herschel, "may be comprehended all which, to *a want of complete, and in most instances, even of partial resolvability* by the power of the 20-feet reflector, unite such a deviation from the circular or elliptic form, or such a want of symmetry (with that form) as preclude their being placed in Class 1, or that of regular nebulæ. This second class comprises many of the most remarkable and interesting objects in the heavens, *as well as the most extensive in respect of the area they occupy.*"

And, referring to this same order of objects, M. Arago says: — "The forms of very large diffuse nebulæ do not appear to admit of definition; they have no regular outline."

Now this coexistence of largeness, irresolvability, irregularity, and indefiniteness of outline, is extremely significant. The fact that the largest nebulæ are either irresolvable or very difficult to resolve, might have been inferred *à priori;* seeing that irresolvability, implying that the aggregation of precipitated matter has gone on to but a small extent, will be found in nebulæ of wide diffusion. Again, the irregularity of these large, irresolvable nebulæ, might also have been expected; seeing that their outlines, compared by Arago to "the fantastic figures which characterize clouds carried away and tossed about by violent and often contrary winds," are similarly characteristic of a mass not yet gathered together by the mutual attraction of its parts. And once more, the fact that these large, irregular, irresolvable nebulæ have indefinite outlines — outlines that fade off insensibly into surrounding darkness — is one of like meaning.

Speaking generally (and of course differences of distance negative anything beyond an average statement), the spiral nebulas are smaller than the irregular nebulæ, and more resolvable; at the same time that they are not so small as the regular nebulæ, and not so resolvable. This is as, according to the hypothesis, it should be. The degree of condensation causing spiral movement, is a degree of condensation also implying masses of flocculi that are larger, and therefore more visible, than those existing in an earlier stage. Moreover, the forms of these spiral nebulæ are quite in harmony with the explanation given. The curves of luminous matter which they exhibit, are *not* such as would be described by more or less discrete masses starting from a state of rest, and moving through a resisting medium to a common centre of gravity; but they *are* such as would be described by masses having their movements modified by the rotation of the medium.

In the centre of a spiral nebula is seen a mass both more luminous and more resolvable than the rest. Assume that, in process of time, all the spiral streaks of luminous matter which converge to this centre are drawn into it, as they must be; assume further, that the flocculi or other discrete bodies constituting these luminous streaks aggregate into larger masses at the same time that they approach the central group, and that the masses forming this central group also aggregate into larger masses (both which are necessary assumptions); and there will finally result a more or less globular group of such larger masses, which will be resolvable with comparative ease. And, as the coalescence and concentration go on, the constituent masses will gradually become fewer, larger, brighter, and more densely collected around the common centre of gravity. See now how completely this inference agrees with observation. "The circular form is that which most commonly characterizes resolvable nebulæ," writes Arago. "Resolvable nebulæ," says Sir John Herschel, "are almost universally round or oval." Moreover, the centre of each group habitually displays a closer clustering of the constituent masses than elsewhere; and it is shown that, under (he law of gravitation, which we know extends to the stars, this distribution is *not* one of equilibrium, but implies progressing concentration. While, just as we inferred that, according to circumstances, the extent to which aggregation has been carried must vary; so we find that, in fact, there are regular nebulæ of all degrees of resolvability, from those consisting of innumerable minute discrete masses, to those in which there are a few large bodies worthy to be called stars.

On the one hand, then, we see that the notion, of late years uncritically received, that the nebulæ are extremely remote galaxies of stars like those which make up our own Milky Way, is totally irreconcileable with the facts — involves us in sundry absurdities. On the other hand, we see that the hypothesis of nebular condensation harmonizes with the most recent results of stellar astronomy: nay more — that it supplies us with an explanation of various appearances which in its absence would be incomprehensible.

Descending now to the Solar System, let us consider first a class of phenomena in some sort transitional — those offered by comets. In comets we have now existing a kind of matter like that out of which, according to the Nebular Hypothesis, the Solar System was evolved. For the explanation of them, we must hence go back to the time when the substances forming the sun and planets were yet unconcentrated.

When diffused matter, precipitated from a rarer medium, is aggregating, there are certain to be here and there produced small flocculi, which, either in consequence of local currents or the conflicting attractions of adjacent masses, remain detached; as do, for instance, minute shreds of cloud in a summer sky. In a concentrating nebula these will, in the great majority of cases, eventually coalesce with the larger flocculi near to them. But it is tolerably evident that some of the remotest of these small flocculi, formed at the outermost parts of the nebula, will *not* coalesce with the larger internal masses, but will slowly follow without overtaking them. The relatively greater resistance of the medium necessitates this. As a single feather falling to the ground will be rapidly left behind by a pillow-full of feathers; so, in their progress to the common centre of gravity, will the outermost shreds of vapour be left behind by the great masses of vapour internally situated. But we are not dependent merely on reasoning for this belief. Observation shows us that the less concentrated external parts of nebulæ, *are* left behind by the more concentrated, internal parts. Examined through high powers, all nebulæ, even when they have assumed regular forms, are seen to be surrounded by luminous streaks, of which the directions show that they are being drawn into the general mass. Still higher powers bring into view still smaller, fainter, and more widely-dispersed streaks. And it cannot be doubted that the minute fragments which no telescopic aid makes visible, are yet more numerous and widely dispersed. Thus far, then, inference and observation are at one.

Granting that the great majority of these outlying portions of nebulous matter will be drawn into the central mass long before it reaches a definite form, the presumption is that some of the very small, far-removed portions will not be so; but that before they arrive near it, the central mass will have contracted into a comparatively moderate bulk. What now will be the characters of these late-arriving portions?

In the first place, they will have extremely eccentric orbits. Left behind at a time when they were moving towards the centre of gravity in slightly-deflected lines, and therefore having but very small angular velocities, they will approach the central mass in greatly elongated ellipses, and rushing round it will go off again into space. That is, they will behave just as we see comets do; whose orbits are usually so eccentric as to be indistinguishable from parabolas.

In the second place, they will come from all parts of the heavens. Our supposition implies that they were left behind at a time when the nebulous mass was of irregular shape, and had not acquired a definite rotary motion; and as the separation of them would not be from any one surface of the nebulous mass more than another the conclusion must be that they will come to the central body from various directions in space. This, too, is exactly what happens. Unlike planets, whose orbits approximate to one plane, comets have orbits that show no relation to each other; but cut the plane of the ecliptic at all angles.

In the third place, applying the reasoning already used, these remotest flocculi of nebulous matter will, at the outset, be deflected from their straight courses to the common centre of gravity, not all on one side, but each on such side as its form determines. And being left behind before the rotation of the nebula is set up, they will severally retain their different individual motions. Hence, following the concentrating mass, they will eventually go round it on all sides; and as often from right to left as from left to right. Here again the inference perfectly corresponds with the facts. While all the planets go round the sun from west to east, comets as often go round the sun from east to west as from west to east. Out of 210 comets known in 1855, 104 are direct, and 106 are retrograde. This equality is what the law of probabilities would indicate.

Then, in the fourth place, the physical constitution of comets completely accords with the hypothesis. The ability of nebulous matter to concentrate into a concrete form, depends on its mass. To bring its ultimate atoms into that proximity requisite for chemical union — requisite, that is, for the production of denser matter — their repulsion must be overcome. The only force antagonistic to their repulsion, is their mutual gravitation. That their mutual gravitation may generate a pressure and temperature of sufficient intensity, there must be an enormous accumulation of them, and even then the approximation can slowly go on only as fast as the evolved heat escapes. But where the quantity of atoms is small, and therefore the force of mutual gravitation small, there will be nothing to coerce the atoms into union. Whence we infer that these detached fragments of nebulous matter will continue in their original state. We find that they do so. Comets consist of an extremely rare medium, which, as shown by the description already quoted from Sir John Herschel, has characters like those we concluded would belong to partially-condensed nebulous matter.

Yet another very significant fact is seen in the distribution of comets. Though they come from all parts of the heavens, they by no means come in equal abundance from all parts of the heavens; but are far more numerous about the poles of the ecliptic than about its plane. Speaking generally, comets having orbit-planes that are highly inclined to the ecliptic, are comets having orbits of which the major axes are highly inclined to the ecliptic — comets that come from high latitudes. This is not a necessary connexion; for the planes of the orbits *might* be highly inclined to the ecliptic while the major axes were inclined to it very little. But in the absence of any habitually-observed relation of this kind, it may safely be concluded that, *on the average,* highly-inclined cometary orbits are cometary orbits with highly-inclined major axes; and that thus, a predominance of cometary orbits cutting the plane of the ecliptic at great angles, implies a predominance of cometary orbits having major axes that cut the ecliptic at great angles. Now the predominance of highly inclined cometary orbits, may be gathered from the following table, compiled by M. Arago, to which we have added a column giving the results up to a date two years later.

Inclinations.	Number of Comets in 1831.	Number of Comets in 1853.	Number of Comets in 1855.
Deg. Deg.			
From 0 to 10	9	19	19
" 10 " 20	13	18	19
" 20 " 30	10	13	14
" 30 " 40	17	22	22
" 40 " 50	14	35	36
" 50 " 60	23	27	29
" 60 " 70	17	23	25
" 70 " 80	19	26	27
" 80 " 90	15	18	19
Total	137	201	210

At first sight this table seems not to warrant our statement. Assuming the alleged general relation between the inclinations of cometary orbits, and the directions in space from which the comets come, the table may be thought to show that the frequency of comets increases as we progress from the plane of the ecliptic up to 45°, and then decreases up to 90°. But this apparent diminution arises from the fact that the successive zones of space rapidly diminish in their areas on approaching the poles. If we allow for this, we shall find that the excess of comets continues to increase up to the highest angles of inclination. In the table below, which, for convenience, is arranged in inverted order, we have taken as standards of comparison the area of the zone round the pole, and the number of comets it contains; and having ascertained the areas of the ether zones, and the numbers of comets they should contain were comets equally distributed, we have shown how great becomes the deficiency in descending from the poles of the ecliptic to its plane.

Between	Area of Zone	Number of Comets, if equally distributed.	Actual number of Comets.	Deficiency.	Relative Abundance.
Deg. Deg.					
90 and 80	1	19	19	0	11·5
80 " 70	2·98	56·6	27	29·6	5·5
70 " 60	4·85	92	25	67	3·12
60 " 50	6·6	125	29	96	2·66
50 " 40	8·13	154	36	118	2·68
40 " 30	9·42	179	22	157	1·4
30 " 20	10·42	198	14	184	0·8
20 " 10	11·1	210	19	191	1·04
10 " 0	11·5	218	19	199	1

In strictness, the calculation should be made with reference, not to the plane of the ecliptic, but to the plane of the sun's equator; and this might or might not render the progression more regular. Probably, too, the progression would be made somewhat different were the calculation based, as it should be, not on the inclinations of orbit-planes, but on the inclinations of major axes. But even as it is, the result is sufficiently significant: since, though the conclusion that comets are 11·5 times more abundant about the poles of the ecliptic than about its plane, can be but a rough approximation to the truth, yet no correction of it is likely very much to change this strong contrast.

What, then, is the meaning of this fact? It has several

meanings. It negatives the supposition, favoured by Laplace among others, that comets are bodies that were wandering in space, or have come from other systems; for the probabilities are infinity to one against the orbits of such wandering bodies showing any definite relation to the plane of the Solar System. For the like reason, it negatives the hypothesis of Lagrange, otherwise objectionable, that comets have resulted from planetary catastrophes analogous to that which is supposed to have produced the asteroids. It clearly shows that, instead of comets being *accidental* members of the Solar System, they are *necessary* members of it — have as distinct a structural relation to it as the planets themselves. That comets are abundant round the axis of the Solar System, and grow rarer as we approach its plane, implies that the genesis of comets has followed some *law* — a law in some way concerned with the genesis of the Solar System.

If we ask for any so-called final cause of this arrangement, none can be assigned: until a probable use for comets has been shown, no reason can be given why they should be thus distributed. But when we consider the question as one of physical science, we see that comets are antithetical to planets, not only in their great rarity, in their motions as indifferently direct or retrograde, in their eccentric orbits, and in the varied directions of those orbits; but we see the antithesis further marked in this, that while planets have some relation to the plane of nebular rotation, comets have some relation to the axis of nebular rotation.[11] And without attempting to explain the nature of this relation, the mere fact that such a relation exists, indicates that comets have resulted from a process of evolution — points to a past time when the matter now forming the Solar System extended to those distant regions of space which comets visit.

See, then, how differently this class of phenomena bears on the antagonistic hypotheses. To the hypothesis commonly received, comets are stumbling-blocks: why there should be hun-

---

11 It is alike remarkable and suggestive, that a parallel relation exists between the distribution of nebulæ and the axis of our galaxy. Just as comets are abundant around the poles of our Solar System, and rare in the neighbourhood of its plane: so are nebulæ abundant around the poles of our sidereal system, and rare in the neighbourhood of its plane.

dreds (or probably thousands) of extremely rare aeriform masses rushing to and fro round the sun, it cannot say; any more than it can explain their physical constitutions, their various and eccentric movements, or their distribution. The hypothesis of evolution, on the other hand, not only allows of the general answer, that they are minor results of the genetic process; but also furnishes us with something like explanations of their several peculiarities.

And now, leaving these erratic bodies, let us turn to the more familiar and important members of the Solar System. It was the remarkable harmony subsisting among their movements, which first made Laplace conceive that the sun, planets, and satellites had resulted from a common genetic process. As Sir William Herschel, by his observations on the nebulæ, was led to the conclusion that stars resulted from the aggregation of diffused matter; so Laplace, by his observations on the structure of the Solar System, was led to the conclusion that only by the rotation of aggregating matter were its peculiarities to be explained. In his "Exposition du Système du Monde," he enumerates as the leading evidences of evolution: — 1. The movements of the planets in the same direction and almost in the same plane; 2. The movements of the satellites in the same direction as those of the planets; 3. The movement of rotation of these various bodies and of the sun in the same direction as the orbitual motions, and in planes little different; 4. The small eccentricity of the orbits of the planets and satellites, as contrasted with the great eccentricity of the cometary orbits. And the probability that these harmonious movements had a common cause, he calculates as two hundred thousand billions to one.

Observe that this immense preponderance of probability does not point to a common cause under the form ordinarily conceived — an Invisible Power working after the method of "a Great Artificer;" but to an Invisible Power working after the method of evolution. For though the supporters of the common hypothesis may argue that it was necessary for the sake of stability that the planets should go round the sun in the same direction and nearly in one plane, they cannot thus account for the direction of the axial motions. The mechanical equilibrium would not have been at all interfered with, had the sun been without any rotatory movement; or had he revolved on his axis in a direction opposite to that in which the planets go round him; or in a direction at right angles to the plane of their orbits. With equal safety the motion of the Moon round the Earth might have been the reverse of the Earth's motion round its axis; or the motion of Jupiter's satellites might similarly have been at variance with his axial motion or that of Saturn's satellites with his. As, however, none of these alternatives have been followed, the uniformity must be considered, in this case as in all others, evidence of subordination to some general law — implies what we call natural causation, as distinguished from arbitrary arrangement.

Hence the hypothesis of evolution would be the only probable one, even in the absence of any clue to the particular mode of evolution. But when we have, propounded by a mathematician whose authority is second to none, a definite theory of this evolution based on established mechanical laws, which accounts for these various peculiarities, as well as for many minor ones, the conclusion that the Solar System *was* evolved becomes almost irresistible.

The general nature of Laplace's theory scarcely needs stating. Books of popular astronomy have familiarized most readers with his conceptions; — namely, that the matter now condensed into the Solar System, once formed a vast rotating spheroid of extreme rarity extending beyond the orbit of Neptune; that as this spheroid contracted, its rate of rotation necessarily increased; that by augmenting centrifugal force its equatorial zone was from time to time prevented from following any further the concentrating mass, and so remained behind as a revolving ring; that each of the revolving rings thus periodically detached, eventually became ruptured at its weakest point, and contracting on itself, gradually aggregated into a rotating mass; that this, like the parent mass, increased in rapidity of rotation as it decreased in size, and, where the centrifugal force was sufficient, similarly threw off rings, which finally collapsed into rotating spheroids; and that thus out of these primary and secondary rings there arose planets and their satellites, while from the central mass there resulted the sun. Moreover, it is tolerably well known that this *à priori* reasoning harmonizes with the results of experiment. Dr. Plateau has shown that when a mass of fluid is, as far may be, protected from the action of external forces, it will, if made to rotate with adequate velocity, form detached rings; and that these rings will break up into spheroids which turn on their axes in the same direction with the central mass. Thus, given the original nebula, which, acquiring a vertical motion in the way we have explained, has at length concentrated into a vast spheroid of aeriform matter moving round its axis — given this, and mechanical principles explain the rest. The genesis of a solar system displaying movements like those observed, maybe predicted; and the reasoning on which the prediction is based is countenanced by experiment.[12]

But now lot us inquire whether, besides these most conspicuous peculiarities of the Solar System, sundry minor ones are not similarly explicable. Take first the relation between the planes of the planetary orbits and the plane of the sun's equator. If, when the nebulous spheroid extended beyond the orbit of Neptune, all parts of it had been revolving exactly in the same plane or rather in parallel planes — if all its parts had had one axis; then the planes of the successive rings would have been coincident with each other and with that of the sun's rotation. But it needs only to go back to the earlier stages of concentration, to see that there could exist no such complete uniformity of motion. The flocculi, already described as precipitated from an irregular and widely-diffused nebula, and as starting from all points to their common centre of gravity, must move not in one plane but in innumerable planes, cutting each other at all angles.

The gradual establishment of a vertical motion such as we

---

[12] It is true that, as expressed by him, these propositions of Laplace are not all beyond dispute. An astronomer of the highest authority, who has favoured me with some criticisms on this essay, alleges that instead of a nebulous ring rupturing at one point, and collapsing into a single mass, "all probability would be in favour of its breaking up into many masses." This alternative result certainly seems to be more likely. But granting that a nebulous ring would break up into many masses, it may still be contended that, since the chances are infinity to one against these being of equal sizes and equidistant, they could not remain evenly distributed round their orbit: this annular chain of gaseous masses would break up into groups of masses; these groups would eventually aggregate into larger groups; and the final result would be the formation of a single mass. I have put the question to an astronomer scarcely second in authority to the one above referred to, and he agrees that this would probably be the process.

saw must eventually arise, and such as we at present see indicated in the spiral nebulæ, is the gradual approach toward motion in one plane — the plane of greatest momentum. But this plane can only slowly become decided. Flocculi not moving in this plane, but entering into the aggregation at various inclinations, will tend to perform their revolutions round its centre in their own planes; and only in course of time will their motions be partly destroyed by conflicting ones, and partly resolved into the general motion. Especially will the outermost portions of the rotating mass retain for long time their more or less independent directions; seeing that neither by friction nor by the central forces will they be so much restrained. Hence the probabilities are, that the planes of the rings first detached will differ considerably from the average plane of the mass; while the planes of those detached latest will differ from it lees. Here, again, inference to a considerable extent agrees with observation. Though the progression is irregular, yet on the average the inclinations decrease on approaching the sun.

Consider next the movements of the planets on their axes. Laplace alleged as one among other evidences of a common genetic cause, that the planets rotate in a direction the same as that in which they go round the sun, and on axes approximately perpendicular to their orbits. Since he wrote, an exception to this general rule has been discovered in the case of Uranus, and another still more recently in the case of Neptune — judging, at least, from the motions of their respective satellites. This anomaly has been thought to throw considerable doubt on his speculation; and at first sight it does so. But a little reflection will, we believe, show that the anomaly is by no means an insoluble one; and that Laplace simply went too far in putting down as a certain result of nebular genesis, what is, in some instances, only a probable result. The cause he pointed out as determining the direction of rotation, is the greater absolute velocity of the outer part of the detached ring. But there are conditions under which this difference of velocity may be relatively insignificant, even if it exists: and others in which, though existing to a considerable extent, it will not suffice to determine the direction of rotation.

Note, in the first place, that in virtue of their origin, the different strata of a concentrating nebulous spheroid, will be very unlikely to move with equal angular velocities: only by friction continued for an indefinite time will their angular velocities be made uniform; and especially will the outermost strata, for reasons just now assigned, maintain for the longest time their differences of movement. Hence, it is possible that in the rings first detached the outer rims may not have greater absolute velocities and thus the resulting planets may have retrograde rotations. Again, the sectional form of the ring is a circumstance of moment; and this form must have differed more or less in every case. To make this clear, some illustration will be necessary. Suppose we take an orange, and assuming the marks of the stalk and the calyx to represent the poles, cut off round the line of the equator a strip of peel. This strip of peel, if placed on the table with its ends meeting, will make a ring shaped like the hoop of a barrel — a ring whose thickness in the line of its diameter is very small, but whose width in a direction perpendicular to its diameter is considerable. Suppose, now, that in place of an orange, which is a spheroid of very slight oblateness, we take a spheroid of very great oblateness, shaped somewhat like a lens of small convexity. If from the edge or equator of this lens-shaped spheroid, a ring of moderate size were cut off, it would be unlike the previous ring in this respect, that its greatest thickness would be in the line of its diameter, and not in a line at right angles to its diameter: it would be a ring shaped somewhat like a quoit, only far more slender. That is to say, according to the oblateness of a rotating spheroid, the detached ring may be either a hoop-shaped ring or a quoit-shaped ring.

One further fact must be noted. In a much-flattened or lens-shaped spheroid, the form of the ring will vary with its bulk. A very slender ring, taking off just the equatorial surface, will be hoop-shaped; while a tolerably massive ring, trenching appreciably on the diameter of the spheroid, will be quoit-shaped. Thus, then, according to the oblateness of the spheroid and the bulkiness of the detached ring, will the greatest thickness of that ring be in the direction of its plane, or in a direction perpendicular to its plane. But this circumstance must greatly affect the rotation of the resulting planet. In a decidedly hoop-shaped nebulous ring, the differences of velocity between the inner and outer surfaces will be very small; and such a ring, aggregating into a mass whose greatest diameter is at right angles to the plane of the orbit, will almost certainly give to this mass a predominant tendency to rotate in a direction at right angles to the plane of the orbit. Where the ring is but little hoop-shaped, and the difference of the inner and outer velocities also greater, as it must be, the opposing tendencies — one to produce rotation in the plane of the orbit, and the other rotation perpendicular to it — will both be influential; and an intermediate plane of rotation will be taken up. While, if the nebulous ring is decidedly quoit-shaped, and therefore aggregates into a mass whose greatest dimension lies in the plane of the orbit, both tendencies will conspire to produce rotation in that plane.

On referring to the facts, we find them, as far as can be judged, in harmony with this view. Considering the enormous circumference of Uranus's orbit, and his comparatively small mass, we may conclude that the ring from which he resulted was a comparatively slender, and therefore a hoop-shaped one: especially if the nebulous mass was at that time less oblate than afterwards, which it must have been. Hence, a plane of rotation nearly perpendicular to his orbit, and a direction of rotation having no reference to his orbitual movement. Saturn has a mass seven times as great, and an orbit of less than half the diameter; whence it follows that his genetic ring, having less than half the circumference, and less than half the vertical thickness (the spheroid being then certainly *as* oblate, and indeed *more* oblate), must have had considerably greater width — must have been less hoop-shaped, and more approaching to the quoit-shaped: notwithstanding difference of density, it must have been at least two or three times as broad in the line of its plane. Consequently, Saturn has a rotatory movement in the same direction as the movement of translation, and in a plane differing from it by thirty degrees only.

In the case of Jupiter, again, whose mass is three and a half times that of Saturn, and whose orbit is little mere than half the size, the genetic ring must, for the like reasons, have been still broader — decidedly quoit-shaped, we may say; and there hence resulted a planet whose plane of rotation differs from that of his orbit by scarcely more than three degrees. Once more, considering the comparative insignificance of Mars, Earth, Venus, and Mercury, it follows that the diminishing circumferences of the rings not sufficing to account for the smallness of the resulting masses, the rings must have been slender ones — must have again approximated to the hoop-shaped; and thus it happens that the planes of rotation again diverge more or less widely from those of the orbits. Taking into account the increasing oblateness of the original spheroid in the successive stages of its concentration, and the different proportions of the detached rings, it seems to us that the respective rotatory motions are not at variance with the hypothesis.

Not only the directions, but also the velocities of rotation are thus explicable. It might naturally be supposed that the large planets would revolve on their axes more slowly than the small ones; our terrestrial experiences incline us to expect this. It is a corollary from the Nebular Hypothesis, however, more especially when interpreted as above, that while large planets will rotate rapidly, small ones will rotate slowly; and we find that in fact they do so. Other things equal, a concentrating nebulous mass that is diffused through a wide space, and whose outer parts have, therefore, to travel from great distances to the common centre of gravity, will acquire a high axial velocity in course of its aggregation: and conversely with a small mass. Still more marked will be the difference where the form of the genetic ring conspires to increase the rate of rotation. Other things equal, a genetic ring that is broadest in the direction of its plane will produce a mass rotating faster than one that is broadest at right angle to its piano and if the ring is absolutely as well as relatively broad, the rotation will be very rapid. These conditions were, as we saw, fulfilled in the case of Jupiter; and Jupiter goes round his axis in less than ten hours. Saturn, in whose case, as above explained, the conditions were less favourable to rapid rotation, takes ten hours and a half. While Mars, Earth, Venus, and Mercury, whose rings must have been slender, take more than double the time: the smallest taking the longest.

From the planets, let us now pass to the satellites. Here, beyond the conspicuous facts commonly adverted to, that they go round their primaries in the same directions that these turn on their axes, in planes diverging but little from their equators, and in orbits nearly circular, there are several significant traits which must not be passed over.

One of them is, that each set of satellites repeats in miniature the relations of the planets to the sun, both in the respects just named, and in the order of the sizes. On progressing from the outside of the Solar System to its centre, we see that there are four large external planets, and four internal ones which are comparatively small. A like contrast holds between the outer and inner satellites in every case. Among the four satellites of Jupiter, the parallel is maintained as well as the comparative smallness of the number allows: the two outer ones are the largest, and the two inner ones the smallest. According to the most recent observations made by Mr. Lassell, the like is true of the four satellites of Uranus. In the case of Saturn, who has eight secondary planets revolving round him, the likeness is still more close in arrangement as in number: the thrice outer satellites are large, the inner ones small; and the contrasts of size are here much greater between the largest, which is nearly as big as Mars, and the smallest, which is with difficulty discovered even by the best telescopes.

Moreover, the analogy does not end here. Just as with the planets, there is at first a general increase of size on travelling inwards from Neptune and Uranus, which do not differ very widely, to Saturn, which is much larger, and to Jupiter, which is the largest; so of the eight satellites of Saturn, the largest is not the outermost, but the outermost save two; so of Jupiter's four secondaries, the largest is the most remote but one. Now these analogies are inexplicable by the theory of final causes. For purposes of lighting, if this be the presumed object of these attendant bodies, it would have been far better had the larger been the nearer: at present, their remoteness renders them of less service than the smallest. To the Nebular Hypothesis, however, these analogies give further support. They show the action of a common physical cause. They imply a *law* of genesis, holding in the secondary systems as in the primary system.

Still more instructive shall we find the distribution of the satellites — their absence in some instances, and their presence in other instances, in smaller or greater numbers. The argument from design fails to account for this distribution. Supposing it be granted that planets nearer the Sun than ourselves, have no need of moons (though, considering that their nights are as dark, and, relatively to their brilliant days, even darker than ours, the need seems quite as great) — supposing this to be granted; what is to be said of Mars, which, placed half as far again from the Sun as we are, has yet no moon? Or again, how are we to explain the fact that Uranus has but half as many moons as Saturn, though he is at double the distance? While, however, the current presumption is untenable, the Nebular Hypothesis furnishes us with an explanation. It actually enables us to predict, by a not very complex calculation, where satellites will be abundant and where they will be absent. The reasoning is as follows.

In a rotating nebulous spheroid that is concentrating into a planet, there are at work two antagonist mechanical tendencies — the centripetal and the centrifugal. While the force of gravitation draws all the atoms of the spheroid together, their tangential momentum is resolvable into two parts, of which one resists gravitation. The ratio which this centrifugal force bears to gravitation, varies, other things equal, as the square of the velocity. Hence, the aggregation of a rotating nebulous spheroid will be more or less strongly opposed by this outward impetus of its particles, according as its rate of rotation is high or low: the opposition, in equal spheroids, being four times as great when the rotation is twice as rapid; nine times as great when it is three times as rapid; and so on. Now, the detachment of a ring from a planet-forming body of nebulous matter, implies that at its equatorial zone the centrifugal force produced by concentration has become so great as to balance gravity. Whence it is tolerably obvious that the detachment of rings will be most frequent from those masses in which the centrifugal tendency bears the greatest ratio to the gravitative tendency. Though it is not possible to calculate what proportions these two tendencies had to each other in the genetic spheroid which produced each planet; it is possible to calculate where each was the greatest and where the least. While it is true that the ratio which centrifugal force now bears to gravity at the equator of each planet, differs widely from that which it bore during the earlier stages of concentration; and while it is true that this change in the ratio, depending on the degree of contraction each planet has undergone, has in no two cases been the same; yet we may fairly conclude that where the ratio is still the greatest, it has been the greatest from the beginning. The satellite-forming tendency which each planet had, will be approximately indicated by the proportion now existing in it between the aggregating power, and the power that has opposed aggregation. On making the requisite calculations, a remarkable harmony with this inference comes out. The following table shows what fraction the centrifugal force is of the centripetal force in every case; and the relation which that fraction bears to the number of satellites.

Mercury.	Venus.	Earth.	Mars.	Jupiter.	Saturn.	Uranus.
$\frac{1}{362}$	$\frac{1}{282}$	$\frac{1}{289}$	$\frac{1}{326}$	$\frac{1}{14}$	$\frac{1}{6{\cdot}2}$	$\frac{1}{9}$
		1 Satellite.		4 Satellites.	8 Satellites and three rings.	4 (or 6 according to Herschel.)

Thus, taking as our standard of comparison the Earth with its one moon, we see that Mercury and Mars, in which the centrifugal force is relatively less, have no moons. Jupiter, in which it is far greater, has four moons. Uranus, in which it is greater still, has certainly four, and probably more than four. Saturn, in which it is the greatest, being nearly one-sixth of gravity, has, including his rings, eleven attendants. The only instance in which there is imperfect conformity with observation is that of Venus. Here it appears that the centrifugal force is relatively a very little greater than in the Earth; and according to the hypothesis, Venus ought, therefore, to have a satellite. Of this seeming anomaly there are two explanations. Not a few astronomers have asserted that Venus *has* a satellite. Cassini, Short, Montaigne of Limoges, Roedkier, and Montbarron, professed to have seen it; and Lambert calculated its elements. Granting, however, that they were mistaken, then is still the fact that the diameter of Venus is variously estimated; and that a very small change in the data would make the fraction less instead of greater than that of the Earth. But admitting the discrepancy, we think that this correspondence, even as it now stands, is one of the strongest confirmations of the Nebular Hypothesis.[13]

---

[13] Since this essay was published, the data of the above calculations have been changed by the discovery that the Sun's distance is three millions of miles less than was supposed. Hence results a diminution in his estimated mass, and in the masses of the planets (except the Earth and Moon). No revised estimate of the masses having yet been published, the table is re-printed in its original form. The diminution of the masses to the alleged extent of about one-tenth, does not essentially alter the relations above pointed out.

Certain more special peculiarities of the satellites must be mentioned as suggestive. One of them is the relation between the period of revolution and that of rotation. No discoverable purpose is served by making the Moon go round its axis in the same time that it goes round the Earth: for our convenience, a more rapid axial motion would have been equally good; and for any possible inhabitants of the Moon, much better. Against the alternative supposition, that the equality occurred by accident, the probabilities are, as Laplace says, infinity to one. But to this arrangement, which is explicable neither as the result of design nor of chance, the Nebular Hypothesis furnishes a clue. In his "Exposition du Système du Monde," Laplace shows, by reasoning too detailed to be here repeated, that under the circumstances such a relation of movements would be likely to establish itself.

Among Jupiter's satellites, which severally display these same synchronous movements, there also exists a still more remarkable relation. "If the mean angular velocity of the first satellite be added to twice that of the third, the sum will be equal to three times that of the second;" and "from this it results that the situations of any two of them being given, that of the third can be found." Now here, as before, no conceivable advantage results. Neither in this case can the connexion have been accidental: the probabilities are infinity to one to the contrary. But again, according to Laplace, the Nebular Hypothesis supplies a solution. Are not these significant facts?

Most significant fact of all, however, is that presented by the rings of Saturn. As Laplace remarks, they are, as it were, still extant witnesses of the genetic process he propounded. Here we have, continuing permanently, forms of matter like those through which each planet and satellite once passed; and their movements are just what, in conformity with the hypothesis, they should be. "La durée de la rotation d'une planete doit done être, d'apres cette hypothèse, plus petite que la durée de la révolution du corps le plus voisin qui circule autour d'elle," says Laplace.[14] And he then points out that the time of Saturn's rotation is to that of his rings as 427 to 438 — an amount of difference such as was to be expected.

But besides the existence of these rings, and their move-

---

[14] "Mécanique Céleste," p. 346.

ments in the required manner, there is a highly suggestive circumstance which Laplace has not remarked — namely, the place of their occurrence. If the Solar System was produced after the manner popularly supposed, then there is no reason why the rings of Saturn should not have encircled him at a comparatively great distance. Or, instead of being given to Saturn, who in their absence would still have had eight satellites, such rings might have been given to Mars, by way of compensation for a moon. Or they might have been given to Uranus, who, for purposes of illumination, has far greater need of them. On the common hypothesis, we repeat, no reason can be assigned for their existence in the place where we find them. But on the hypothesis of evolution, the arrangement, so far from offering a difficulty, offers another confirmation. These rings are found where alone they could have been produced — close to the body of a planet whose centrifugal force bears a great proportion to his gravitative force. That permanent rings should exist at any great distance from a planet's body, is, on the Nebular Hypothesis, manifestly impossible. Kings detached early in the process of concentration, and therefore consisting of gaseous matter having extremely little power of cohesion, can have no ability to resist the disrupting forces due to imperfect balance; and must, therefore, collapse into satellites. A liquid ring is the only one admitting of permanence. But a liquid ring can be produced only when the aggregation is approaching its extreme — only when gaseous matter is passing into liquid, and the mass is about to assume the planetary form. And even then it cannot be produced save under special conditions. Gaining a rapidly-increasing preponderance, as the gravitative force does during the closing stages of concentration, the centrifugal force cannot in ordinary cases cause the detachment of rings when the mass has become dense. Only where the centrifugal force has all along been very great, and remains powerful to the last, as in Saturn, can liquid rings be formed. Thus the Nebular Hypothesis shows us why such appendages surround Saturn, but exist nowhere else.

And then, let us not forget the fact, discovered within these few years, that Saturn possesses a *nebulous* ring, through which his body is seen as through a thick veil. In a position where alone such a thing seems preservable — suspended, as it were, between the denser rings and the planet — there still continues one of these annular masses of diffused matter from which satellites and planets are believed to have originated.

We find, then, that besides those most conspicuous peculiarities of the Solar System, which first suggested the theory of its evolution, there are many minor ones pointing in the same direction. Were there no other evidence, these mechanical arrangements would, considered in their totality, go far to establish the Nebular Hypothesis.

From the mechanical arrangements of the Solar System, turn we now to its physical characters; and, first, let us consider the inferences deducible from relative specific gravities.

The fact that, speaking generally, the denser planets are the nearer to the Sun, is by some considered as adding another to the many indications of nebular origin. Legitimately assuming that the outermost parts of a rotating nebulous spheroid, in its earlier stages of concentration, will be comparatively rare; and that the increasing density which the whole mass acquires as it contracts, must hold of the outermost parts as well as the rest; it is argued that the rings successively detached will be more and more dense, and will form planets of higher and higher specific gravities. But passing over other objections, this explanation is quite inadequate to account for the facts. Using the Earth as a standard of comparison, the relative densities run thus: —

Neptune	Uranus	Saturn	Jupiter	Mars	Earth	Venus	Mercury	Sun
0·14	0·24	0·14	0·24	0·95	1·00	0·92	1·12	0·25

Two seemingly insurmountable objections are presented by this series. The first is, that the progression is but a broken one. Neptune is as dense as Saturn, which, by the hypothesis, it ought not to be. Uranus is as dense as Jupiter, which it ought not to be. Uranus is denser than Saturn, and the Earth is denser than Venus — facts which not only give no countenance to, but directly contradict, the alleged explanation. The second objection, still more manifestly fatal, is the low specific gravity of the Sun. If, when the matter of the Sun filled the orbit of Mercury, its state of aggregation was such that the detached ring formed a planet having a specific gravity equal to that of iron; then the Sun itself, now that it has concentrated, should have a specific gravity much greater than that of iron; whereas its specific gravity is not much above that of water. Instead of being far denser than the nearest planet, it is not one-fourth as dense.

While these anomalies render untenable the position that the relative specific gravities of the planets are direct indications of nebular condensation; it by no means follows that they negative it. When all the factors are taken in account, there appears to be an interpretation consistent with the hypothesis of Laplace.

Several causes may be assigned for the unlike specific gravities of the members of the Solar-system: — 1. Differences among them in respect of the elementary substances composing them; or in the proportions of such elementary substances, if they contain the same kinds. 2. Differences among them in respect of the quantities of matter they contain; for, other things equal, the mutual gravitation of molecules will make a larger mass denser than a smaller. 3. Differences among their temperatures; for, other things equal, those having higher temperatures will have lower specific gravities. 4. Differences among their structures in respect of the relative amounts of solid, liquid, and gaseous matters composing them. Of these conceivable causes, that commonly assigned is the first, more or less modified by the second. The extremely low specific gravity of Saturn, which but little exceeds that of cork, is supposed to arise from the intrinsic lightness of his substance: the tacit assumption being that the atmosphere of Saturn incloses a globe solid like our own, or otherwise, as Whewell supposes, liquid. That the Sun weighs not much more than an equal bulk of water, is taken as evidence that the specific gravity of the matter he consists of, is but little greater than that of water; although, considering his enormous gravitative force, which at his surface is twenty-eight times the gravitative force at the surface of the Earth, and considering the pressure which his immense mass (315,511 times that of the Earth), must exert upon his central parts; the matter he is made of must, in such case, be widely unlike the matters we know. Nevertheless, the current hypothesis is, that the Sun and planets are either solid or liquid, or have solid crusts with liquid nuclei: their unlike specific gravities resulting from unlikenesses of nature in their component materials.[15]

---

[15] Excepting some alterations rendering its statements more specific, the above paragraph stands as it did when originally written five-

But now ignoring teleological implications, and recognizing the truth disclosed by spectrum-analysis since the foregoing passage was originally written (the truth that the same elements exist in the Sun and in the Earth, and probably in the other planets, since even in the stars sundry of them have been identified) — recognizing thus the untenability of the assumption that the great differences between the specific gravities of the planets can be due to diversities among their component matters; we are left with the alternative of supposing that these great differences result from differences in the conditions under which their component matters exist — differences specified under the heads 2, 3, and 4. Of these remaining possible causes, that numbered 2 we must at once reject; for supposing the materials the same, and all other circumstances the same, superiority in mass would involve greater density: Jupiter and Saturn would have higher specific gravities than the Earth and Venus, instead of having far lower specific gravities. Thus, as tenable causes for these differences of specific gravity, we are restricted to differences of temperature, and differences in the proportions between solid, liquid, and gaseous matters; which last, indeed, we may regard as accompanying differences of temperature. But these causes are just the causes inferable from the nebular hypothesis; as we shall see on comparing the processes of concentration and cooling in large and small spheroids of nebulous matter.

That large masses cool more slowly than small ones, is a truth illustrated at every fire-side by the rapid blackening of little cinders which fall into the ashes, in contrast with the long-continued redness of big lumps. There are several reasons for such contrasts. One is the relation between increase of surface and increase of content: surfaces, in similar bodies, increasing as the squares of the dimensions while contents increase as their cubes. Hence, contrasting the Earth and Jupiter, whose diameter is about eleven times that of the Earth, it results that while his surface is 121 times as great, his content is 1331 times as great. Now even (supposing we begin with like temperatures and like densities) if the only effect was that through a given area of surface eleven times more matter had to be cooled in the one case

---

and-twenty years ago. What follows is so far changed as to bring its arguments into harmony with discoveries since made.

than in the other, there would be a vast difference between the times occupied in concentration. But the difference would be much greater than that consequent on these geometrical relations. The escape of heat from a cooling mass is effected by conduction, or by convection, or by both. In a solid it is wholly by conduction; in a liquid or gas the chief part is played by convection — by circulating currents which continually transpose the hotter and cooler parts. Now in fluid spheroids — gaseous, or liquid, or mixed — increasing size entails an increasing obstacle to cooling, because of the increasing distances to be travelled by the circulating currents. Of course the relation is not a simple one: the velocities of the currents will doubtless be unlike. But it is manifest that in a sphere of eleven times the diameter, the transit of matter from centre to surface and back from surface to centre, will take a much longer time; even if its movement is unrestrained. In a rotating spheroid, however, there come into play forces which restrain its movement in a degree augmenting with the velocity of rotation. In such a spheroid the respective portions of matter (supposing them equal in their angular velocities round the axis, which they will tend more and more to become as the density increases) must vary in their absolute velocities, according to their distances from the axis; and each portion, having a momentum varying with its velocity, cannot have its distance from the axis changed by circulating currents, which it must be in greater or less degree, without loss or gain in its momentum: through the medium of fluid friction, force must be expended, now in increasing its motion and now in retarding its motion. Hence, when the larger spheroid has also a higher velocity of rotation — when, as in the case of Jupiter and the Earth, the angular velocity of the larger body is more than twice that of the smaller; the relative slowness of the circulating currents, and the consequent retardation of cooling, must be much greater than is implied by the extra distances to be travelled. There are, however, certain factors which, working in an opposite way, qualify these effects. Other things equal, mutual gravitation among the parts of a large mass will cause a greater evolution of heat than is similarly caused in a small mass; and the resulting difference of temperature, as well as the greater contrast between the temperatures of inner and outer parts in the larger mass than in the smaller, will tend to produce more rapid dissipation of heat. To this must be added the greater velocity of the circulating currents which the intenser forces at work in larger spheroids will produce — a contrast made still greater by the relatively smaller retardation by friction to which the more voluminous currents are exposed. In these causes we may recognize a possible explanation of the otherwise anomalous fact that the Sun, though having a thousand times the mass of Jupiter, has yet reached as advanced a stage of concentration; for the force of gravity in the Sun, which at his surface is some ten times that at the surface of Jupiter, must expose his central parts to a pressure relatively very intense; producing during contraction a relatively rapid genesis of heat. And it is further to be remarked, that though the circulating currents in the Sun have far greater distances to travel, yet since his rotation is relatively so slow that the angular velocity of his substance is but about one-sixtieth of that of Jupiter's substance, the resulting obstacle to circulating currents is relatively small, and the escape of heat far less retarded. Here, too, we may note that in the co-operation of these factors, there seems a reason for the greater concentration reached by Jupiter than by Saturn, though Saturn is the elder as well as the smaller of the two; for at the same time that the gravitative force in Jupiter is much greater than that in Saturn, his velocity of rotation is less.

But now, not judging more than roughly of the effects of these several factors, co-operating in various ways and degrees, some to aid concentration and others to resist it, it is sufficiently manifest that, other things equal, the larger nebulous spheroids, longer in losing their heat, will more slowly reach high specific gravities; and that where the contrasts in size are so immense as those between the greater and the smaller planets, the smaller may have reached relatively high specific gravities when the greater have reached but relatively low ones. Further, it appears that such qualification of the process as results from the more rapid genesis of heat in the larger masses, will be countervailed where high velocity of rotation greatly impedes the circulating currents. Thus interpreted then, the various specific gravities of the planets do not conflict with the nebular hypothesis, but may be held to furnish further evidences of it. And now observe that the evidences thus furnished *à priori,* appear to be in course of verification *à posteriori.* In his work entitled *Saturn and its System,* Mr. Proctor has brought together proofs of sundry kinds that

Jupiter and Saturn, having vaporous atmospheres of vast depths, subject to immense perturbations, have not reached any such stages of concentration and cooling as have been reached by the smaller planets.

On pursuing further the above considerations, we reach certain conclusions respecting the internal constitutions of the planets, which suggest possible solutions of difficulties. In conformity with the teleological view above referred to, it has been taken for granted that each member of the Solar System is either solid or has a solid shell filled with molten matter: the Earth, supposed by some to have one of these structures and by some to have the other, being regarded as typical of the rest. Here, in the first place, it should be observed that neither of these is more than an hypothesis; and we must not let the familiarity of either, or apparent probability of either, shut out the consideration of a third hypothesis if it should be physically possible. This third hypothesis is that the central part of every celestial body consists of gaseous matter — that the aeriform state which originally characterized the entire mass is permanently retained by its nucleus; and that around this there has, during evolution, been formed a liquid shell, the outer surface of which has in some cases eventually solidified.

"But what," it may be asked, "will become of this gaseous nucleus when exposed to the enormous gravitative pressure of a shell some thousands of miles thick? How can aeriform matter withstand such a pressure?" Very readily. It has been proved that, even when the heat generated by compression is allowed to escape, some gases remain uncondensible by any force we can produce. An unsuccessful attempt lately made at Vienna to liquify oxygen, clearly shows this enormous resistance. The steel piston employed was literally shortened by the pressure used; and yet the gas remained unliquified! If, then, the expansive force is thus immense when the heat evolved is dissipated, what must it be when that heat is in great measure detained, as in the case we are considering? Indeed the experiences of M. Cagniard de Latour have shown that gases may, under pressure, acquire the density of liquids while retaining the aeriform state, provided the temperature continues extremely high. In such a case every addition to the heat is an addition to the repulsive power of the atoms: the increased pressure itself generates an increased ability to resist; and this remains true to whatever extent the compression is carried. Indeed it is a corollary from the persistence of force, that if, under increasing pressure, a gas retains all the heat evolved, its resisting force is *absolutely unlimited.* Hence the internal planetary structure we have described is as physically stable a one as that commonly assumed.[16]

And now, concluding that this hypothesis, not negatived by established scientific truths, may reasonably be discussed, let us consider what grounds there are for entertaining it as probable. Let us consider what likelihood there is that a constitution of this kind would be assumed by a concentrating spheroid of nebulous matter. Throughout such a spheroid there will be convection-currents. Taking for granted that there are present several elements (conventionally so-called) which have possibly, or probably, been formed during earlier stages of nebular concentration; and bearing in mind that these elements in a gaseous state have different degrees of liquitiability, we may infer that there must come a time when the currents of mixed gases ascending from the hotter and denser interior will, on approaching the outside, arrive at a region in which, losing part of their heat by expansion, and a further part by radiation through the superjacent gases, they will begin to precipitate their more condensible components into vapour. That is to say, there will result an internal space filled with perfectly gaseous matters, a surrounding layer in which some of these hold in suspension others which have taken the form of cloud, and an outermost layer consisting of the least condensible gases. Where there are sundry elements having different condensation-points this structure is likely to become further involved. The cloudy layer will possibly — may we not say probably? — become a compound one: certain of the more condensible elements passing into the vaporous form before they have reached the outer surface of it and forming an innermost cloudy layer of denser kind. If, now, we pursue in thought this process of concentration — if we recognize increasing mutual

---

[16] This paragraph stands just as it did when published in *The Westminster Review* for July, 1858, when the evidences it assigns were the best which could be given. In what follows I have utilized the results of more recent researches; and especially those of Prof. Andrews concerning the "critical point" of gases.

gravitation of the parts, increasing pressure of the outer layers upon the inner, increasing genesis and radiation of heat, along with more rapid convection-currents and an ever-thickening cloudy layer — we must infer that eventually the inner surface of this layer will arrive at what is known as "the critical point" of one of the elements, or of the mixture: a point at which the specific gravities of the gaseous and liquid forms are the same. That is to say, there will arise an innermost stratum throughout which this element or mixture exists indifferently as gas or liquid — now with slight change of pressure or temperature assuming the liquid form, and now, with some counter change passing back into the gaseous form. We may infer that such stratum will long remain in this indeterminate state. A liquid at or near the critical point, expands or contracts greatly as the conditions are altered; so that there tends continually to arise a mingling of portions having refractive powers suggesting a liquid form of aggregation with other portions having refractive powers suggesting a gaseous form. But we may reasonably conclude that eventually there will result a continuous liquid stratum. Though perpetually perturbed and broken by currents, this liquid stratum will perpetually restore itself. Of greater specific gravity than the vaporous strata above it; of equal specific gravity with the contained gases immediately in contact with it; and of less specific gravity than the deeper portions of the gaseous spheroid these form; it will fulfil the conditions to mechanical stability. Once established it will thicken as fast as radiation of the planet's heat into space permits. Evidently the inner surface of this shell must maintain itself at the critical point; since such cooling of all the outer structures as permits the temperature of this surface to fall, implies some contraction of these outer structures with accompanying approach to the common centre of gravity; some increase of gravitative force involving higher pressure upon the contained gaseous spheroid, and consequently, some rise of temperature: divergence from the critical point either way will be instantly checked. And then observe, finally and chiefly, that the matter contained within this molten shell must retain permanently its gaseous form. For the law is that at a temperature above its critical point, a gas cannot by any amount of pressure be reduced to its correlative liquid; and the temperature throughout this gaseous spheroid must everywhere be higher than at the surface, where loss of its heat is going on. Should it be objected to this argument that, at the critical temperature, a substance in its liquid state is more compressible than in its gaseous state, and that, consequently, the liquid when formed, acquiring a higher specific gravity, would sink towards the centre, and thus the distribution described would be unstable; the reply is, that though, wore one clement only concerned, this might be the case, there is reason to think it would not be the case if there were present the mingled gases of several elements, some of less specific gravity than others while more condensible.

When in course of time this shell has become thick, the outer surface, ever radiating its heat into space, will fall very much below the inner in temperature. Chemical combination between the superjacent oxygen and the bases, metallic and other, forming the periphery of the molten layer, will take place; and with a further lapse of time and fall of temperature, solidified portions will float upon its surface as do the scoriæ on a lava stream. That is to say, the solidification of the outside of the molten shell, will be analogous to the preceding liquifaction of the outside of the gaseous nucleus.

I am aware that this hypothesis is incongruous with conclusions drawn by certain distinguished physicists. Sir William Thomson originally indorsed the inference of Mr. Hopkins, that the Earth's crust must be at least 800 miles thick; but now thinks this inference invalid. He has contended, too, that a solid crust upon a molten globe could not sustain itself, since, when fissured, its fragments would sink; but this argument is met by the allegation that iron, copper, brass, whinstone and granite, are less dense when solid than when liquid, and unquestionably bismuth is so.[17] But Sir William Thomson still holds it certain that the

[17] Even could no such demurrer be entered, it might be contended that Sir William Thomson's conclusion is invalid. In this, as in other cases, he has inferred positive results from questionable data. The inference that the crust, breaking into fragments, would sink into the subjacent molten matter, assumes that the two are homogeneous in composition. The assumption is not simply gratuitous; it is extremely improbable. Most likely before any solids were formed, there would have arisen some such segregation of the molten elements as brought to the surface a larger proportion of the lighter ones—silicon, aluminium, calcium, sodium, etc. But even without supposing this, it is manifest

substance of the Earth is extremely rigid. He contends that were it otherwise, the tidal wave produced in its substance would cause rises and falls of the outer surface, corresponding with those formed in the ocean; and that there would consequently be no visible tide. Hence he infers that the Earth, though once molten, is now solid all through; and that its remaining heat is escaping by conduction only. In his elaborate work on the *Physics of the Earth's Crust,* the Rev. O. Fisher argues that the hypothesis of solidity fails to account for the geological facts. He also calculates that the contraction in that case resulting from loss of heat, would cause corrugations of the surface not approaching in amount to those which actually exist: assuming the existence of a liquid substratum, the inference he draws from observed temperature beneath the Alps, being that the solid crust is about 25 miles thick. But now, returning to the difficulty raised by Sir William Thomson, there arises the question — What modification results if, instead of a globe consisting wholly of molten matter inclosed in a solid crust, we suppose this crust to be the cooled and solidified covering of a molten shell (perhaps of no great thickness) containing matter which, though as dense as this shell at their surface of contact and increasing in density towards its centre, maintains its gaseous form, and instead of being relatively inelastic and viscous is highly elastic and mobile? What would be the character of the tidal wave in a relatively thin shell, thus supported by a gaseous nucleus which resisted with great promptness its deformation? What would be the character of the tidal wave in the gaseous nucleus itself? And then, further, what would be the respective retardations of the tidal wave in the gaseous nucleus, in the viscous molten shell, and in the superjacent water? May we not infer that the positions in longitude of the waves in the nucleus, and in the shell, would differ so widely

---

that oxides of these elements, stable at high temperature, would be formed at the surface of contact between the molten matter and the superjacent atmosphere, largely charged with oxygen; and that, being lighter and less fusible than the elements themselves, they would float upon its surface in virtue of their absolutely smaller specific gravity, irrespective of their state of aggregation. The fact that the Earth's crust, mainly composed of these oxides, has a specific gravity 2, in contrast with the average specific gravity of the Earth, which is more than 5, harmonizes with this conclusion.

as partially to cancel one another, and thus leave the wave in the water partially uncancelled? An additional reason for asking whether the tidal phenomena may not thus be rendered accountable, may be given. The hypothesis of entire solidity which Mr. Fisher gives reasons for rejecting as untenable, and the hypothesis which he himself thinks necessitated, that the matter contained within the Earth's crust is molten near its surface but, by Sir W. Thomson's argument, solid towards its centre, both assume central solidity on the strength of the fact that solids subject to pressure have higher melting points than when free; and that therefore, conversely, liquids may be solidified by pressure. But is not solidification by pressure dependent on escape of the contained heat? Will it be alleged that a quantity of metal in a molten state can by any pressure be reduced to a solid state, if none of its heat is suffered to escape? Unless this be alleged, it must be admitted that the matter of a molten globe could not solidify at the centre, where increase of pressure generated its correlative increase of temperature. Contrariwise it may be reasonably inferred that just as when raised to its critical point of temperature, a gas cannot be reduced to the liquid form by any amount of pressure; so, when a liquid retains all that heat which made it liquid, no amount of pressure will solidify it. But, of course, if the foregoing reasoning is accepted, which implies that liquifaction did not begin at the centre but near the surface, the hypothesis of solidification at the centre is excluded.

Indirect evidence supporting the general view above set forth, is furnished by the asteroids or planetoids. The hypothesis of Olbers, propounded when only four of them were known, that they are fragments of an exploded planet which once occupied the region they fill, is generally held less probable than that of Laplace, that they were formed from a nebulous ring which separated into several parts instead of collapsing into a single mass. But this latter hypothesis, which looked tenable at the time when only four planetoids had been discovered, does not look so tenable now that the number discovered reaches 230: the implication being, moreover, that other hundreds remain to be discovered. Against this hypothesis it may be urged that did a nebulous ring break up into numerous small portions, revolving round the Sun with approximately equal velocities, the annular series of them would inevitably have some point of least attraction between its adjacent members at which parting would take place, followed by collapse of its members upon one another till a single body was formed. Moreover, their mean distances from the Sun could scarcely differ so much that some are twice others: it could hardly happen that the annular space included between their most unlike mean distances would be more than 100 millions of miles across, and that the space occupied by their widest excursions would be 270 millions of miles across. Again, the parts of such a ring could not well have orbits much inclined to one another, or much inclined to the average plane of the solar system, as are sundry of these planetoid orbits. Further, their orbits could not differ greatly in eccentricity as they do — one, if not more of them, to the extent of cutting the orbit of Mars. Surely no portion of an outer nebulous ring could thus intrude upon the region of an inner nebulous ring, at the same time that other portions almost intruded upon the region of a remoter nebulous ring. Once more, there could not arise any considerable differences between the times in which the discrete portions of such a ring revolved round the sun; to the extent of some being thrice others. But all these traits of the planetoids, incongruous with the supposition of Laplace, are congruous with that of Olbers; and are, indeed, necessitated by it. Fragments propelled in all directions by an explosion, multitudinous in number and various in size, would inevitably acquire orbits differing greatly in their periodic times, their inclinations, and their eccentricities. The inference drawn from the present derangement of their orbits, that if the planetoids once formed parts of one mass, it must have exploded myriads of years ago, is no difficulty. Moreover, we have at the same time an adequate cause for meteors, whether solitary or in periodic swarms; since, along with the larger fragments there would be shot into space immeasurably greater numbers of smaller fragments, themselves of all sizes, some in groups approximately the same in direction and velocity, and others singly — fragments, the physical characters of which would bear testimony to their having once belonged to some great mass where active physical and chemical processes had been going on.[18]

---

[18] While these revised pages are standing in type, it has occurred to me that evidence for or against the hypothesis of Olbers ought to be

furnished by the assemblage of planetoids considered in respect of their sizes and distribution. If they resulted from the bursting of a planet once revolving in the region they occupy, the implications are: — first, that the fragments must be most abundant in the space immediately about the original orbit, and less abundant far away from it; second, that the large fragments must be relatively few, while of smaller fragments the numbers will increase as the sizes decrease; third, that as some among the smaller fragments will be propelled further than any of the larger, the widest deviations in mean distance from the mean distance of the original planet, will be presented by the smallest members of the assemblage; and fourth, that the orbits differing most from the rest in eccentricity and in inclination will be among those of these smallest members. With a view to verification, I first referred to a recent general account of the planetoids in the *Annuaire de l'Observatoire Royal de Bruxelles,* 1881, 48e. *Année.* Comparison of the extremes tended to show that some such relations exist. Taking the distance from the Earth to the Sun as 1, there is a region lying between 2·55 and 2·8 where the planetoids are found in maximum abundance. The mean distance between these extremes, 2·66, does not greatly differ from the average of the distances of the four largest planetoids, 2·64. May we not say that the thick clustering of these bodies about this distance (which is, however, rather less than that assigned for the planet by Bode's empirical law) in contrast with the wide scattering of the few whose distances are little more than 2, or exceed 3, is a fact answering to anticipation? Again, among those having measured sizes, we may note that the orbits of the two largest, Vesta and Ceres, having diameters respectively of 420 and 338 kilometres, have eccentricities falling between ·05 and ·1; while the orbits of the two smallest, Menippe and Eva, 19 and 22 kilometres in diameter respectively, have eccentricities falling between ·25 and ·3, and between ·3 and ·35. And then among those more recently discovered, having diameters so small that they have not yet been measured, come the extremely erratic ones — Hilda, which with a mean distance of nearly 4 approaches the orbit of Jupiter, Æthra, having an orbit so eccentric that it cuts the orbit of Mars, and Medusa, which comes nearest to the Sun. Systematic comparisons yield results only approximately correct, because the numerous planetoids remaining to be discovered, will greatly increase those groups which now include the smallest; for, of the added members the average magnitude is each year less. Allowing for this, we find evidence to the point. Analysis of the tables at the end of *Chambers's Descriptive Astronomy,* shows that there is only one planetoid between the sixth and seventh star-magnitudes. Between the seventh and eighth star-magnitudes there is also but one. Between the eighth and ninth there are six. Between the ninth and tenth there are sixteen. Between the tenth and eleventh there are thirty-six. Of smaller magnitudes than the eleventh there are sixty-one specified; and the forty odd more recently discovered, having unspecified star-magnitudes, are presumably among these smallest. To which must be added the sixty more that have been found since the date of the table. Thus it is clear that with decrease in the sizes of those bodies, there goes increase in their numbers. Kindred evidence is furnished if we broadly contrast their mean distances. Out of the 14 largest planetoids having star-magnitudes falling between 6·5 and 9·5 there is not one having a mean distance that exceeds 3. Of those having magnitudes between 9·5 and 10 there are 10; and of these but one has a mean distance greater than 3. Of those between 10 and 10·5 there are sixteen; and of these also there is but one exceeding 3 in mean distance. The group of next smallest in magnitude contains 20, and of these 5 have this great mean distance. In the next group there are 22, and of these 9 have this great mean distance. The next group, 27, contains 5 such; the next, 9, contains 2 such; the next, 7, contains 2 such; and the single planetoid forming the next group has itself this great mean distance. Hence the ratios to their respective groups of those having this great mean distance run thus:—0·0, 0·1, 0·06, 0·25, 0·4, 0·19, 0·22, 0·29, 1·0. Were there to be added the 100 additional ones, this progression might not improbably become more regular—a conclusion supported by the fact that out of the last 9 included at the end of the table in Newcomb's *Popular Astronomy,* which have not their elements given by Chambers, there are 4 having this great mean distance. If the average eccentricities of the orbits of the planetoids grouped according to their decreasing sizes are compared, no very definite results are disclosed, excepting this, that Polyhymnia, Eurydice, Atalanta, and Æthra, which have the greatest eccentricities (falling between ·3 and ·38) are all among those of smallest star-magnitudes. Nor when we consider the inclinations of the orbits do we meet with obvious verification; since the proportion of highly-inclined orbits among the smaller planetoids, does not appear to be greater than among the others. But consideration shows that there are two ways in which these last comparisons are vitiated. One is, that the inclinations are measured from the plane of the ecliptic instead of being measured from the plane of the orbit of the hypothetical planet. The other, and more important one, is, that the search for planetoids has naturally been carried on in that comparatively narrow zone within which most of their orbits fall; and that, consequently, those having the most highly-inclined orbits are the least likely to have been detected; especially if they are at the same time among the smallest. Moreover, considering the general relation between the inclinations of planetoid orbits and their eccentricities, it is

Taking Olbers' supposition, then, as the most tenable one, let us ask how such an explosion could have occurred. If planets are internally constituted as is commonly assumed, no conceivable cause for it can be named. A solid mass may crack and fall to pieces, but it cannot violently explode. So, too, with a liquid mass covered by a crust. Though, if contained in an unyielding shell and artificially raised to a very high temperature, a liquid might burst the shell and simultaneously flash into vapour, yet, if contained in a yielding crust, like that of a planet, it would not do so. But the planetary structure above supposed, supplies the requisite conditions to an explosion, and an adequate cause for it. We have in the interior of the mass, a cavity serving as a sufficient reservoir of force. We have this cavity filled with gaseous matters of high tension. We have in the chemical affinities of these matters a source of enormous expansive power capable of being suddenly liberated. And we have in the increasing heat consequent on progressing concentration, a cause of such instantaneous chemical change and the resulting catastrophe. Thus, as yielding a conceivable origin for these innumerable erratic bodies, varying in their sizes from those of small planets down to those of fragments an ounce or two in weight, which, though members of the Solar System, differ so greatly from the rest in their behaviour, while also differing multitudinously among themselves, the hypothesis gains indirect support.

In the preceding sections a good deal has been said, directly or by implication, concerning the heat resulting from the concentration of nebulous matter into planets. But quite apart from the speculations set forth in them, there is to be noted the fact that in the present conditions of celestial bodies at large, with respect to temperature, we find additional materials for building up the argument; and these, too, of the most substantial character.

Heat must inevitably be generated by the aggregation of diffused matter into a concrete form; and throughout our reasonings we have assumed that such generation of heat has been an ac-

---

probable that among the orbits of these undetected planetoids are many of the most eccentric. But while recognizing the incompleteness of the evidence, it seems to me that it goes far to justify the hypothesis of Olbers, and is quite incongruous with that of Laplace.

companiment of nebular condensation. If, then, the Nebular Hypothesis be true, we ought to find in all the heavenly bodies, either present high temperatures or marks of past high temperatures.

As far as observation can reach, the facts prove to be what theory requires. Various evidences conspire to show that, below a certain depth, the Earth is still molten. And that it was once wholly molten, is implied by the circumstance that the rate at which the temperature increases on descending below its surface, is such as would be found in a mass that had been cooling for an indefinite period. The Moon, too, shows us, by its corrugations and its conspicuous volcanoes, that in it there has been a process of refrigeration and contraction, like that which has gone on in the Earth. And in Venus, the existence of mountains similarly indicates the shrinking of a solidifying crust, or an igneous reaction of the interior upon it, or both.

On the common theory of creation, these phenomena are inexplicable. To what end the Earth should once have existed in a molten state, incapable of supporting life, it cannot say. To satisfy this supposition, the Earth should have been originally created in a state fit for the assumed purposes of creation; and similarly with the other planets. While, therefore, to the Nebular Hypothesis the evidence of original incandescence and still continued internal heat, furnish strong confirmation, they are, to the antagonist hypothesis, insurmountable difficulties.

But the argument from temperature does not end here. There remains to be noticed a more conspicuous and still more significant fact. If the Solar System was formed by the concentration of diffused matter, which evolved heat while gravitating into its present dense form; then there are certain obvious corollaries respecting the relative temperatures of the resulting bodies. Other things equal, the latest-formed mass will be the latest in cooling — will, for an almost infinite time, possess a greater heat than the earlier-formed ones. Other things equal, the largest mass will, because of its superior aggregative force, become hotter than the others, and radiate more intensely. Other things equal, the largest mass, notwithstanding the higher temperature it reaches, will, in consequence of its relatively small surface, be the slowest in losing its evolved heat. And hence, if there is one mass which was not only formed after the rest, but exceeds them enormously in size, it follows that this one will reach an intensity of incandescence much beyond that reached by the rest; and will continue in a state of intense incandescence long after the rest have cooled.

Such a mass we have in the Sun. It is a corollary from the Nebular Hypothesis, that the matter forming the Sun assumed its present concrete form, at a period much more recent than that at which the planets became definite bodies. The quantity of matter contained in the Sun is nearly five million times that contained in the smallest planet, and above a thousand times that contained in the largest. And while, from the enormous gravitative force of the atoms, the evolution of heat has been intense, the facilities of radiation have been relatively small. Hence the still-continued high temperature. Just that condition of the central body which is a necessary inference from the Nebular Hypothesis, we find actually existing in the Sun.

It may be well to consider a little more closely, what is the probable condition of the Sun's surface. Round the globe of incandescent molten substances, thus conceived to form the visible body of the Sun, there is known to exist a voluminous atmosphere: the inferior brilliancy of the Sun's border, and the appearances during a total eclipse, alike show this.[19] What now must be the constitution of this atmosphere? At a temperature approaching a thousand times that of molten iron, which is the calculated temperature of the solar surface, very many, if not all, of the substances we know as solid, would become gaseous; and though the Sun's enormous attractive force must be a powerful check on this tendency to assume the form of vapour, yet it cannot be questioned that if the body of the Sun consists of molten substances, some of them must be constantly undergoing evaporation. That the dense gases thus continually being generated will form the entire mass of the solar atmosphere, is not probable. If anything is to be inferred, either from the Nebular Hypothesis, or from the analogies supplied by the planets, it must be concluded that the outermost part of the solar atmosphere consists of what are called permanent gases — gases that are not condensible into fluid even at low temperatures. If we consider what must have been the state of things here, when the surface of the Earth was

---

[19] See Herschel's "Outlines of Asnronomy."

molten, we shall see that round the still molten surface of the Sun, there probably exists a stratum of dense aeriform matter, made up of sublimed metals and metallic compounds, and above this a stratum of comparatively rare medium analogous to air. What now will happen with these two strata? Did they both consist of permanent gases, they could not remain separate: according to a well-known law, they would eventually form a homogeneous mixture. But this will by no means happen when the lower stratum consists of matters that are gaseous only at excessively high temperatures. Given off from a molten surface, ascending, expanding, and cooling, these will presently reach a limit of elevation above which they cannot exist as vapour, but must condense and precipitate. Meanwhile the upper stratum, habitually charged with its quantum of these denser matters, as our air with its quantum of water, and ready to deposit them on any depression of temperature, must be habitually unable to take up any more of the lower stratum; and therefore this lower stratum will remain quite distinct from it.

Since the foregoing paragraph was originally published, in 1858, the proposition it enunciates as a corollary from the Nebular Hypothesis, has been in great part verified. The marvellous disclosures made by spectrum-analysis, have proved beyond the possibility of doubt, that the solar atmosphere contains, in a gaseous state, the metals, iron, calcium, magnesium, sodium, chromium, and nickel, along with small quantities of barium, copper, and zinc. That there exist in the solar atmosphere other metals like those which we have on the Earth, is probable; and that it contains elements which are unknown to us, is very possible.

Be this as it may, however, the proposition that the Sun's atmosphere consists largely of metallic vapours, must take rank as an established truth; and that the incandescent body of the Sun consists of molten metals, follows almost of necessity. That an *à priori* inference which probably seemed to many readers wildly speculative, should be thus conclusively justified by observations, made without reference to any theory, is a striking fact; and it gives yet further support to the hypothesis from which this *à priori* conclusion was drawn. It may be well to add that Kirchhoff, to whom we owe this discovery respecting the constitution of the solar atmosphere, himself remarks in his memoir of 1861, that the facts disclosed are in harmony with the Nebular Hypothesis.

And here let us not omit to note also, the significant bearing which Kirchhoff's results have on the doctrine contended for in a foregoing section. Leaving out the barium, copper, and zinc, of which the quantities are inferred to be small, the metals existing as vapours in the Sun's atmosphere, and by consequence as molten in his incandescent body, have an average specific gravity of 4·25. But the average specific gravity of the Sun is about 1. How is this discrepancy to be explained? To say that the Sun consists almost wholly of the three lighter metals named, would be quite unwarranted by the evidence: the results of spectrum-analysis would just as much warrant the assertion that the Sun consists almost wholly of the three heavier. Three metals (two of them heavy) having been already left out of the estimate because their quantities appear to be small, the only legitimate assumption on which to base an estimate of specific gravity, is that the rest are present in something like equal amounts. Is it then that the lighter metals exist in larger proportions in the molten mass, though not in the atmosphere? This is very unlikely: the known habitudes of matter rather imply that the reverse is the case. Is it then that under the conditions of temperature and gravitation existing in the Sun, the state of liquid aggregation is wholly unlike that existing here? This is a very strong assumption: it is one for which our terrestrial experiences afford no adequate war rant; and if such unlikeness exists, it is very improbable that it should produce so immense a contrast in specific gravity as that of 4 to 1. The more legitimate conclusion is that the Sun's body is not made up of molten matter all through; but that it consists of a molten shell with a gaseous nucleus. And this we have seen to be a corollary from the Nebular Hypothesis.

Considered in their *ensemble,* the several groups of evidences assigned amount almost to proof. We have seen that, when critically examined, the speculations of late years current respecting the nature of the nebulæ, commit their promulgators to sundry absurdities; while, on the other hand, we see that the various appearances these nebulæ present, are explicable as different stages in the precipitation and aggregation of diffused matter. We find that comets, alike by their physical constitution, their immensely-elongated and variously-directed orbits, the distribution of those orbits, and their manifest structural relation to the Solar System, bear testimony to the past existence of that system in a nebulous form. Not only do those obvious peculiarities in the motions of the planets which first suggested the Nebular Hypothesis, supply proofs of it, but on closer examination we discover, in the slightly-diverging inclinations of their orbits, in their various rates of rotation, and their differently-directed axes of rotation, that the planets yield us yet further testimony; while the satellites, by sundry traits, and especially by their occurrence in greater or less abundance where the hypothesis implies greater or less abundance, confirm this testimony. By tracing out the process of planetary condensation, we are led to conclusions respecting the internal structure of planets which at once explain their anomalous specific gravities, and at the same time reconcile various seemingly contradictory facts. Once more, it turns out that what is *à priori* inferable from the Nebular Hypothesis respecting the temperatures of the resulting bodies, is just what observation establishes; and that both the absolute and the relative temperatures of the Sun and planets are thus accounted for. When we contemplate these various evidences in their totality — when we observe that, by the Nebular Hypothesis, the leading phenomena of the Solar System, and the heavens in general, are explicable; and when, on the other hand, we consider that the current cosmogony is not only without a single fact to stand on, but is at variance with all our positive knowledge of Nature; we see that the proof becomes overwhelming.

It remains only to point out that while the genesis of the Solar System, and of countless other systems like it, is thus rendered comprehensible, the ultimate mystery continues as great as ever. The problem of existence is not solved: it is simply removed further back. The Nebular Hypothesis throws no light on the origin of diffused matter; and diffused matter as much needs accounting for as concrete matter. The genesis of an atom is not easier to conceive than the genesis of a planet. Nay, indeed, so far from making the Universe a less mystery than before, it makes it a greater mystery. Creation by manufacture is a much lower thing than creation by evolution. A man can put together a machine; but he cannot make a machine develop itself. The ingenious artizan, able as some have been, so far to imitate vitality as to produce a mechanical pianoforte-player, may in some sort conceive how, by greater skill, a complete man might be artificially produced; but he is unable to conceive how such a complex organism gradually arises out of a minute structureless germ. That our harmonious universe once existed potentially as formless diffused matter, and has slowly grown into its present organized state, is a far more astonishing fact than would have been its formation after the artificial method vulgarly supposed. Those who hold it legitimate to argue from phenomena to noumena, may rightly contend that the Nebular Hypothesis implies a First Cause as much transcending "the mechanical God of Paley," as this does the fetish of the savage.

Note: — The argument on page 251, omits a factor which is of greater importance than that assigned. A concentrating nebula will necessarily have a central region which is denser than its peripheral regions. Manifestly, any flocculus moving towards this central region, will have its course affected not only by the unequal reactions of the medium on its unsymmetrically disposed parts, but also by differences of pressure consequent on inequalities of density of the medium on its two sides, especially near the nucleus: unless its line of movement is directly towards the centre of mass, against which the chances are infinity to one, it must be made to diverge. Hence, such motion as it has acquired, directed not towards the common centre of gravity but towards one side of it, will be changed into a motion of revolution round that centre. And since the momenta of the flocculi thus drawn in from various directions on different sides, are certain not to cancel one another completely, the outstanding effect in some one direction will become a general motion of rotation.

Here, as a not unnatural sequence, I am led to point out the invalidity of the reason assigned by Mons. Babinet for rejection of the Nebular Hypothesis. He has calculated that taking the existing Sun, with its observed angular velocity, its substance, if expanded so as to fill the orbit of Neptune, would have nothing approaching the angular velocity which the time of revolution of that planet implies. The assumption he makes is inadmissible. He supposes that all parts of the nebulous spheroid when it filled Neptune's orbit, had the same angular velocity. But the process of nebular condensation as indicated on page 251 and above, implies that the remoter flocculi of nebulous matter, later in reaching the central mass, and forming its peripheral portions, will acquire during their longer journeys towards it greater velocities. An inspection of one of the spiral nebulæ, as 51st or 99th Messier, at once shows that the outlying portions when they reach the nucleus, will form an equatorial belt moving round the common centre more rapidly than the rest. Thus the central parts will have small angular velocities, while there will be increasing angular velocities of parts increasingly remote from the centre. And while the density of the spheroid continues small, fluid friction will scarcely at all change these differences.

VII
BAIN ON THE EMOTIONS AND THE WILL

After the controversy between the Neptunists and the Vulcanists had been long carried on without definite results, there came a reaction against all speculative geology. Reasoning without adequate data having led to nothing, inquirers went into the opposite extreme, and confining themselves wholly to collecting data, relinquished reasoning. The Geological Society of London was formed with the express object of accumulating evidence; for many years hypotheses were forbidden at its meetings; and only of late have attempts to organize the mass of observations into consistent theory been tolerated.

This reaction and subsequent re-reaction, well illustrate the recent history of English thought in general. The time was when our countrymen speculated, certainly to as great an extent as any other people, on all those high questions which present themselves to the human intellect; and, indeed, a glance at the systems of philosophy that are or have been current on the Continent, suffices to show how much other nations owe to the discoveries of our ancestors. For a generation or two, however, these more abstract subjects have fallen into neglect; and, among those who plume themselves on being "practical," even into contempt. Partly, perhaps, a natural accompaniment of our rapid material growth, this intellectual phase has been in great measure due to the exhaustion of argument, and the necessity for better data. Not so much with a conscious recognition of the end to be subserved, as from an unconscious subordination to that rhythm traceable in social changes as in other things, an era of theorizing without observing, has been followed by an era of observing without theorizing. During the long-continued devotion to concrete science, an immense quantity of raw material for abstract science has been accumulated; and now there is obviously commencing a period in which this accumulated raw material will be organized into consistent theory. On all sides — equally in the inorganic sciences, in the science of life, and in the science of society — may we note the tendency to pass from the superficial and empirical to the more profound and rational.

In Psychology this change is conspicuous. The facts brought to light by anatomists and physiologists during the last fifty years, are at length being used towards the interpretation of this highest class of biological phenomena; and already there is promise of a great advance. The work of Mr. Alexander Bain, of which the second volume has been recently issued, may be regarded as especially characteristic of the transition. It gives us in orderly arrangement, the great mass of evidence supplied by modern science towards the building-up of a coherent system of mental philosophy. It is not in itself a system of mental philosophy, properly so called; but a classified collection of materials for such a system, presented with that method and insight which scientific discipline generates, and accompanied with occasional passages of an analytical character. It is indeed that which it in the main professes to be — a natural history of the mind.

Were we to say that the researches of the naturalist who collects and dissects and describes species, bear the same relation to the researches of the comparative anatomist tracing out the laws of organization, which Mr. Bain's labours bear to the labours of the abstract psychologist, we should be going somewhat too far; for Mr. Bain's work is not wholly descriptive. Still, however, such an analogy conveys the best general conception of what he has done; and serves most clearly to indicate its needfulness. For as, before there can be made anything like true generalizations respecting the classification of organisms and the laws of organization, there must be an extensive accumulation of the facts presented in numerous organic bodies; so, without a tolerably-complete delineation of mental phenomena of all orders, there can scarcely arise any adequate theory of the mind. Until recently, mental science has been pursued much as physical science was pursued by the ancients: not by drawing conclusions from observations and experiments, but by drawing them from arbitrary *à priori* assumptions. This course, long since abandoned in the one case with immense advantage, is gradually being abandoned in the other; and the treatment of Psychology as a division of natural history, shows that the abandonment will soon be complete.

Estimated as a means to higher results, Mr. Bain's work is of great value. Of its kind it is the most scientific in conception, the most catholic in spirit, and the most complete in execution. Besides delineating the various classes of mental phenomena as seen under that stronger light thrown on them by modern science, it includes in the picture much which previous writers had omitted — partly from prejudice, partly from ignorance. We refer more especially to the participation of bodily organs in mental changes; and the addition to the primary mental changes, of those many secondary ones which the actions of the bodily organs generate. Mr. Bain has, we believe, been the first to appreciate the importance of this element in our states of consciousness; and it is one of his merits that he shows how constant and large an element it is. Further, the relations of voluntary and involuntary movements are elucidated in a way that was not possible to writers unacquainted with the modern doctrine of reflex action. And beyond this, some of the analytical passages that here and there occur, contain important ideas.

Valuable, however, as is Mr. Bain's work, we regard it as essentially transitional. It presents in a digested form the results of a period of observation; adds to these results many well-delineated facts collected by himself; arranges new and old materials with that more scientific method which the discipline of our times has fostered; and so prepare the way for better generalizations. But almost of necessity its classifications and conclusions are provisional. In the growth of each science, not only is correct observation needful for the formation of true theory; but true theory is needful as a preliminary to correct observation. Of course we do not intend this assertion to be taken literally; but as a strong expression of the fact that the two must advance hand in hand. The first crude theory or rough classification, based on very slight knowledge of the phenomena, is requisite as a means of reducing the phenomena to some kind of order; and as supplying a conception with which fresh phenomena may be compared, and their agreement or disagreement noted. Incongruities being by and by made manifest by wider examination of cases, there comes such modification of the theory as brings it into a nearer correspondence with the evidence. This reacts to the further advance of observation. More extensive and complete observation brings additional corrections of theory. And so on till the truth is reached. In mental science, the systematic collection of facts having but recently commenced, it is not to be expected that the results can be at once rightly formulated All that may be looked for are approximate generalizations which will presently serve for the better directing of inquiry. Hence, even were it not now possible to say in what way it does so, we might be tolerably certain that Mr. Bain's work bears the stamp of the inchoate state of Psychology.

We think, however, that it will not be difficult to find in what respects its organization is provisional; and at the same time to show what must be the nature of a more complete organization. We propose here to attempt this: illustrating our positions from his recently-issued second volume.

Is it possible to make a true classification without the aid of analysis? or must there not be an analytical basis to every true classification? Can the real relations of things be determined by the obvious characteristics of the things? or does it not commonly happen that certain hidden characteristics, on which the obvious ones depend, are the truly significant ones? This is the preliminary question which a glance at Mr. Bain's scheme of the emotions suggests.

Though not avowedly, yet by implication, Mr. Bain assumes that a right conception of the nature, the order, and the relations of the emotions, may be arrived at by contemplating their conspicuous objective and subjective characters, as displayed in the adult. After pointing out that we lack those means of classification which serve in the case of the sensations, he says —

"In these circumstances we must turn our attention to *the manner of diffusion* of the different passions and emotions, in order to obtain a basis of classification analogous to the arrangement of the sensations. If what we have already advanced on that subject be at all well founded, this is the genuine turning point of the method to be chosen, for the same mode of diffusion will always be accompanied by the same mental experience, and each of the two aspects would identify, and would be evidence of, the other. There is, therefore, nothing so thoroughly characteristic of any state of feeling as the nature of the diffusive wave that embodies it, or the various organs specially roused into action by it, together with the manner of the action. The only drawback is our comparative ignorance, and our inability to discern the precise character of the diffusive currents in every case; a radical imperfection in the science of mind as constituted at present.

"Our own consciousness, formerly reckoned the only medium of knowledge to the mental philosopher, must therefore be still referred to as a principal moans of discriminating the varieties of human feeling. We have the power of noting agreement and difference among our conscious states, and on this we can raise a structure of classification. We recognise such generalities as pleasure, pain, love, anger, through the property of mental or intellectual discrimination that accompanies in our mind the fact of an emotion. A certain degree of precision is attainable by this mode of mental comparison and analysis; the farther we can carry such precision the better; but that is no reason why it should stand alone to the neglect of the corporeal embodiments through which one mind reveals itself to others. The companionship of inward feeling with bodily manifestation is a fact of the human constitution, and deserves to be studied as such; and it would be difficult to find a place more appropriate than a treatise on the mind for setting forth the conjunctions and sequences traceable in this department of nature. I shall make no scruple in conjoining with the description of the mental phenomena the physical appearances, in so far as I am able to ascertain them.

"There is still one other quarter to be referred to in settling a complete arrangement of the emotions, namely, the varieties of human conduct, and the machinery created in subservience to our common susceptibilities. For example, the vast superstructure of fine art has its foundations in human feeling, and in rendering an account of this we are led to recognise the interesting group of artistic or æsthetic emotions. The same outward reference to conduct and creations brings to light the so-called moral sense in man, whose foundations in the mental system have accordingly to be examined.

"Combining together these various indications, or sources of discrimination, — outward objects, diffusive mode or expression, inward consciousness, resulting conduct and institutions — I adopt the following arrangement of the families or natural orders of emotion."

Here, then, are confessedly adopted, as bases of classification, the most manifest characters of the emotions; as discerned subjectively, and objectively. The mode of diffusion of an emotion is one of its outside aspects; the institutions it generates form another of its outside aspects; and though the peculiarities of the emotion as a state of consciousness, seem to express its intrinsic and ultimate nature, yet such peculiarities as are perceptible by simple introspection, must also be classed as superficial peculiarities. It is a familiar fact that various intellectual states of consciousness turn out, when analyzed, to have natures widely unlike those which at first appear; and we believe the like will prove true of emotional states of consciousness. Just as our concept of space, which is apt to be thought a simple, undecomposable concept, is yet resolvable into experiences quite different from that state of consciousness which we call space; so, probably, the sentiment of affection or reverence is compounded of elements that are severally distinct from the whole which they make up. And much as a classification of our ideas which dealt with the idea of space as though it were ultimate, would be a classification of ideas by their externals; so, a classification of our emotions, which, regarding them as simple, describes their aspects in ordinary consciousness, is a classification of emotions by their externals.

Thus, then, Mr. Bain's grouping is throughout determined by the most manifest attributes — those objectively displayed in the natural language of the emotions, and in the social phenomena that result from them, and those subjectively displayed in the aspects the emotions assume in an analytical consciousness. And the question is — Can they be correctly grouped after this method?

We think not; and had Mr. Bain carried farther an idea with which he has set out, he would probably have seen that they cannot. As already said, he avowedly adopts "the natural-history-method:" not only referring to it in his preface, but in his first chapter giving examples of botanical and zoological classifications, as illustrating the mode in which he proposes to deal with the emotions. This we conceive to be a philosophical conception; and we have only to regret that Mr. Bain has overlooked some of its most important implications. For in what has essentially consisted the progress of natural-history-classification? In the abandonment of grouping by external, conspicuous characters; and in the making of certain internal, but all-essential characters, the bases of groups. Whales are not now ranged along with fish, because in their general forms and habits of life they resemble fish; but they are ranged with mammals, because the type of their organization, as ascertained by dissection, corresponds with that of the mammals. No longer considered as sea-weeds in virtue of their forms and modes of growth, zoophytes are now shown, by examination of their economy, to belong to the animal kingdom.

It is found, then, that the discovery of real relationships involves analysis. It has turned out that the earlier classifications, guided by general resemblances, though containing much truth, and though very useful provisionally, were yet in many cases radically wrong; and that the true affinities of organisms, and the true homologies of their parts, are to be made out only by examining their hidden structures. Another fact of great significance in the history of classification is also to be noted. Very frequently the kinship of an organism cannot be made out even by exhaustive analysis, if that analysis is confined to the adult structure. In many cases it is needful to examine the structure in its earlier stages; and even in its embryonic stage. So difficult was it, for instance, to determine the true position of the Cirrhipedia among animals, by examining mature individuals only, that Cuvier erroneously classed them with Mollusca, even after dissecting them; and not until their early forms were discovered, were they clearly proved to belong to the Crustacea. So important, indeed, is the study of development as a means to classification, that the first zoologists now hold it to be the only absolute criterion.

Here, then, in the advance of natural-history-classification, are two fundamental facts, which should be borne in mind when classifying the emotions. If, as Mr. Bain rightly assumes, the emotions are to be grouped after the natural-history-method; then it should be the natural history-method in its complete form, and not in its rude form. Mr. Bain will doubtless agree in the position, that a correct account of the emotions in their natures and relations, must correspond with a correct account of the nervous system — must form another side of the same ultimate facts. Structure and function must necessarily harmonize. Structures which have with each other certain ultimate connexions, must have functions that have answering connexions. Structures that have arisen in certain ways, must have functions that have arisen in parallel ways. And hence if analysis and development are needful for the right interpretation of structures, they must be needful for the right interpretation of functions. Just as a scientific description of the digestive organs, must include not only their obvious forms and connexions, but their microscopic characters, and also the ways in which they severally result by differentiation from the primitive mucous membrane; so must a scientific account of the nervous system, include its general arrangements, its minute structure, and its mode of evolution; and so must a scientific account of nervous actions, include the answering three elements. Alike in classing separate organisms, and in classing the parts of the same organism, the complete natural-history-method involves ultimate analysis, aided by development; and Mr. Bain, in not basing his classification of the emotions on characters reached through these aids, has fallen short of the conception with which he set out.

"But," it will perhaps be asked, "how are the emotions to be analyzed, and their modes of evolution to be ascertained? Different animals, and different organs of the same animal, may readily be compared in their internal and microscopic structures, as also in their developments; but functions, and especially such functions as the emotions, do not admit of like comparisons."

It must be admitted that the application of these methods is here by no means so easy. Though we can note differences and similarities between the internal formations of two animals; it is difficult to contrast the mental states of two animals. Though the true morphological relations of organs may be made out by the observations of embryos; yet, where such organs are inactive before birth, we cannot completely trace the history of their actions. Obviously, too, the pursuance of inquiries of the kind indicated, raises questions which science is not yet prepared to answer; as, for instance — Whether all nervous functions, in common with all other functions, arise by gradual differentiations, as their organs do? Whether the emotions are, therefore, to be regarded as divergent modes of action, that have become unlike by successive modifications? Whether, as two organs which originally budded out of the same membrane, have not only become different as they developed, but have also severally become compound internally, though externally simple: so two emotions, simple and near akin in their roots, may not only have grown unlike, but may also have grown involved in their natures, though seeming homogeneous to consciousness. And here, indeed, in the inability of existing science to answer these questions which underlie a true psychological classification, we see how purely provisional any present classification is likely to be.

Nevertheless, even now, classification may be aided by development and ultimate analysis to a considerable extent; and the defect in Mr. Bain's work is, that he has not systematically availed himself of them as far as possible. Thus we may, in the first place, study the evolution of the emotions up through the various grades of the animal kingdom: observing which of them are earliest and exist with the lowest organization and intelligence; in what order the others accompany higher endowments; and how they are severally related to the conditions of life. In the second place, we may note the emotional differences between the lower and the higher human races — may regard as earlier and simpler those feelings which are common to both, and as later and more compound those which are characteristic of the most civilized. In the third place, we may observe the order in which the emotions unfold during the progress from infancy to maturity. And lastly, comparing these three kinds of emotional development, displayed in the ascending grades of the animal kingdom, in the advance of the civilized races, and in individual history, we may see in what respects they harmonize, and what are the implied general truths.

Having gathered together and generalized these several classes of facts, analysis of the emotions would be made easier. Setting out with the unquestionable assumption, that every new form of emotion making its appearance in the individual or the race, is a modification of some pre-existing emotion, or a compounding of several pre-existing emotions; we should be greatly aided by knowing what always are the pre-existing emotions. When, for example, we find that very few if any of the lower animals show any love of accumulation, and that this feeling is absent in infancy — when we see that an infant in arms exhibits anger, fear, wonder, while yet it manifests no desire of permanent possession, and that a brute which has no acquisitive emotion can nevertheless feel attachment, jealousy, love of approbation; we may suspect that the feeling which property satisfies, is compounded out of simpler and deeper feelings. We may conclude that as, when a dog hides a bone, there must exist in him a prospective gratification of hunger; so there must similarly at first, in all cases where anything is secured or taken possession of, exist an ideal excitement of the feeling which that thing will gratify. We may further conclude that when the intelligence is such that a variety of objects come to be utilized for different purposes — when, as among savages, divers wants are satisfied through the articles appropriated for weapons, shelter, clothing, ornament; the act of appropriating comes to be one constantly involving agreeable associations, and one which is therefore pleasurable, irrespective of the end subserved. And when, as in civilized life, the property acquired is of a kind not conducing to one order of gratifications, but is capable of administering to all gratifications, the pleasure of acquiring property grows more distinct from each of the various pleasures subserved — is more completely differentiated into a separate emotion.

This illustration, roughly as it is sketched, will show what we mean by the use of comparative psychology in aid of classification. Ascertaining by induction the actual order of evolution of the emotions, we are led to suspect this to be their order of successive dependence; and are so led to recognize their order of ascending complexity; and by consequence their true groupings.

Thus, in the very process of arranging the emotions into grades, beginning with those involved in the lowest forms of conscious activity and end with those peculiar to the adult civilized man, the way is opened for that ultimate analysis which alone can lead us to the true science of the matter. For when we find both that there exist in a man feelings which do not exist in a child, and that the European is characterized by some sentiments which are wholly or in a great part absent from the savage — when we see that, besides the new emotions that arise spontaneously as the individual becomes completely organized, there are new emotions making their appearance in the more advanced divisions of our race; we are led to ask — How are new emotions generated? The lowest savages have not even the ideas of justice or mercy: they have neither words for them nor can they be made to conceive them: and the manifestation of them by Europeans they ascribe to fear or cunning. There are æsthetic emotions common among ourselves, that are scarcely in any degree experienced by some inferior races; as, for instance, those produced by music. To which instances may be added the less marked but more numerous contrasts that exist between civilized races in the degrees of their several emotions. And if it is manifest, both that all the emotions are capable of being permanently modified in the course of successive generations, and that what must be classed as new emotions may be brought into existence; then it follows that nothing like a true conception of the emotions is to be obtained, until we understand how they are evolved.

Comparative psychology, while it raises this inquiry, prepares the way for answering it. When observing the differences between races, we can scarcely fail to observe also how these differences correspond with differences in their conditions of existence, and therefore in their daily experiences. Note the contrast between the circumstances and between the emotional natures of savage and civilized. Among be lowest races of men, love of property stimulates to the obtainment only of such things as satisfy immediate desires or desires of the immediate future. Improvidence is the rule: there is little effort to meet remote contingencies. But the growth of established societies, having gradually given security of possession, there has been an increasing tendency to provide for coming years: there has been a constant exercise of the feeling which is satisfied by a provision for the future; and there has been a growth of this feeling so great that it now prompts accumulation to an extent beyond what is needful. Note, again, that under the discipline of social life — under a comparative abstinence from aggressive actions, and a performance of those mutually-serviceable actions implied by the division of labour — there has been a development of those gentle emotions of which inferior races exhibit but the rudiments. Savages delight in giving pain rather than pleasure — are almost devoid of sympathy. While among ourselves philanthropy organizes itself in laws, establishes numerous institutions, and dictates countless private benefactions.

From which and other like facts, does it not seem an unavoidable inference that new emotions are developed by new experiences — new habits of life? All are familiar with the truth, that in the individual, each feeling maybe strengthened by performing those actions which it prompts; and to say that the feeling is *strengthened,* is to say that it is in part *made* by these actions. We know further, that not unfrequently, individuals, by persistence in special courses of conduct, acquire special likings for such courses disagreeable as these may be to others; and these whims, or morbid tastes, imply incipient emotions corresponding to these special activities. We know that emotional characteristics, in common with all others, are hereditary; and the differences between civilized nations descended from the same stock, show us the cumulative results of small modifications hereditarily transmitted. And when we see that between savage and civilized races, which diverged from each other in the remote past, and have for a hundred generations followed modes of life becoming ever more unlike, there exist still greater emotional contrasts; may we not infer that the more or less distinct emotions which characterize civilized races, are the organized results of certain daily-repeated combinations of mental states which social life involves? Must we not say that habits not only modify emotions in the individual, and not only beget tendencies to like habits and accompanying emotions in descendants, but that when the conditions of the race make the habits persistent, this progressive modification may go on to the extent of producing emotions so far distinct as to seem new? And if so, we may suspect that such new emotions, and by implication all emotions analytically considered, consist of aggregated and consolidated groups of those simpler feelings which habitually occur together in experience: that they result from combined experiences, and are constituted of them.

When, in the circumstances of any race, some one kind of action or set of actions, sensation or set of sensations, is usually followed, or accompanied by, various other sets of actions or sensations, and so entails a large mass of pleasurable or painful states of consciousness; these, by frequent repetition, become so connected together that the initial action or sensation brings the ideas of all the rest crowding into consciousness: producing, in a degree, the pleasures or pains that have before been felt in reality. And when this relation, besides being frequently repeated in the individual, occurs in successive generations, all the many nervous actions involved tend to grow organically connected. They become incipiently reflex; and on the occurrence of the appropriate stimulus, the whole nervous apparatus which in past generations was brought into activity by this stimulus, becomes nascently excited. Even while yet there have been no individual experiences, a vague feeling of pleasure or pain is produced; constituting what we may call the body of the emotion. And when the experiences of past generations come to be repeated in the individual, the emotion gains both strength and definiteness; and is accompanied by the appropriate specific ideas.

This view of the matter, which we believe the established truths of Physiology and Psychology unite in indicating, and which is the view that generalizes the phenomena of habit, of national characteristics, of civilization in its moral aspects, at the same time that it gives us a conception of emotion in its origin and ultimate nature, may be illustrated from the mental modifications undergone by animals.

It is well-known that on newly-discovered lands not inhabited by man, birds are so devoid of fear as to allow themselves to be knocked over with sticks; but that in the course of generations, they acquire such a dread of man as to fly on his approach; and that this dread is manifested by young as well as old. Now unless this change be ascribed to the killing-off of the least fearful, and the preservation and multiplication of the more fearful, which, considering the comparatively small number killed by man, is an inadequate cause; it must be ascribed to accumulated experiences; and each experience must be held to have a share in producing it. We must conclude that in each bird that escapes with injuries inflicted by man, or is alarmed by the outcries of other members of the flock (gregarious creatures of any intelligence being necessarily more or less sympathetic), there is established an association of ideas between the human aspect and the pains, direct and indirect, suffered from human agency. And we must further conclude, that the state of consciousness which impels the bird to take flight, is at first nothing more than an ideal reproduction of those painful impressions which before followed man's approach; that such ideal reproduction becomes more vivid and more massive as the painful experiences, direct or sympathetic, increase; and that thus the emotion in its incipient state, is nothing else than an aggregation of the revived pains before experienced.

As, in the course of generations, the young birds of this race begin to display a fear of man before yet they have been injured by him; it is an unavoidable inference that the nervous system of the race has been organically modified by these experiences: we have no choice but to conclude that when a young bird is thus led to fly, it is because the impression produced on its senses by the approaching man, entails, through an incipiently-reflex action, a partial excitement of all those nerves which in its ancestors had been excited under the like conditions; that this partial excitement has its accompanying painful consciousness; and that the vague painful consciousness thus arising, constitutes emotion proper — *emotion undecomposable into specific experiences, and therefore seemingly homogeneous.*

If such be the explanation of the fact in this case, then it is in all cases. If emotion is so generated here, then it is so generated throughout. We must perforce conclude that the emotional modifications displayed by different nations, and those higher emotions by which civilized are distinguished from savage, are to be accounted for on the same principle. And concluding this, we are led strongly to suspect that the emotions in general have severally thus originated.

Perhaps we have now made sufficiently clear what we mean by the study of the emotions through analysis and development. We have aimed to justify the positions that, without analysis aided by development, there cannot be a true natural history of the emotions; and that a natural history of the emotions based on external characters, can be but provisional. We think that Mr. Bain, in confining himself to an account of the emotions as they exist in the adult civilized man, has neglected those classes of facts out of which the science of the matter must chiefly be built. It is true that he has treated of habits as modifying emotions in the individual; but he has not recognized the fact, that where conditions render habits persistent in successive generations, such modifications are cumulative: he has not hinted that the modifications produced by habit are emotions in the making. It is true, also, that he occasionally refers to the characteristics of children; but he does not systematically trace the changes through which childhood passes into manhood, as throwing light on the order and genesis of the emotions. It is further true that he here and there refers to national traits in illustration of his subject; but these stand as isolated facts, having no general significance: there is no hint of any relation between them and the national circumstances; while all those many moral contrasts between lower and higher races which throw great light on classification, are passed over. And once more, it is true that many passages of his work, and sometimes, indeed, whole sections of it, are analytical; but his analyses are incidental — they do not underlie his entire scheme, but are here and there added to it. In brief, he has written a Descriptive Psychology, which does not appeal to Comparative Psychology and Analytical Psychology for its leading ideas. And in doing this, he has omitted much that should be included in a natural history of the mind; while to that part of the subject with which he has dealt, he has given a necessarily-imperfect organization.

Even leaving out of view the absence of those methods and criteria on which we have been insisting, it appears to us that meritorious as is Mr. Bain's book in its details, it is defective in some of its leading ideas. The first paragraphs of his first chapter, quite startled us by the strangeness of their definitions — a strangeness which can scarcely be ascribed to laxity of expression. The paragraphs run thus: —

"Mind is comprised under three heads — Emotion, Volition, and Intellect.

"Emotion is the name here used to comprehend all that is understood by feelings, states of feeling, pleasures, pains, passions, sentiments, affections. Consciousness, and conscious states also for the most part denote modes of emotion, although there is such a thing as the Intellectual consciousness.

"Volition, on the other hand, indicates the great fact that our Pleasures and Pains, which are not the whole of our emotions, prompt us to action, or stimulate the active machinery of the living framework to perform such operations as procure the first and abate the last. To withdraw from a scalding heat and cling to a gentle warmth, are exercises of volition."

The last of these definitions, which we may most conveniently take first, seems to us very faulty. We cannot but feel astonished that Mr. Bain, familiar as he is with the phenomena of reflex action, should have so expressed himself as to include a great part of them along with the phenomena of volition. He seems to be ignoring the discriminations of modern science, and returning to the vague conceptions of the past — nay more, he is comprehending under volition what even the popular speech would hardly bring under it. If you were to blame any one for snatching his foot from the scalding water into which he had inadvertently put it, he would tell you that he could not help it; and his reply would be indorsed by the general experience, that the withdrawal of a limb from contact with something extremely hot, is quite involuntary — that it takes place not only without volition, but in defiance of an effort of will to maintain the contact. How, then, can that be instanced as an example of volition, which occurs even when volition is antagonistic? We are quite aware that it is impossible to draw any absolute line of demarcation between automatic actions and actions which are not automatic. Doubtless we may pass gradually from the purely reflex, through the consensual, to the voluntary. Taking the case Mr. Bain cites, it is manifest that from a heat of such moderate degree that the withdrawal from it is wholly voluntary, we may advance by infinitesimal steps to a heat which compels involuntary withdrawal; and that there is a stage at which the voluntary and involuntary actions are mixed. But the difficulty of absolute discrimination is no reason for neglecting the broad general contrast; any more than it is for confounding light with darkness. If we are to include as examples of volition, all cases in which pleasures and pains "stimulate the active machinery of the living framework to perform such operations as procure the first and abate the last," then we must consider sneezing and coughing, as examples of volition; and Mr. Bain surely cannot mean this. Indeed, we must confess ourselves at a loss. On the one hand if he does not mean it, his expression is lax to a degree that surprises us in so careful a writer. On the other hand, if he does mean it, we cannot understand his point of view.

A parallel criticism applies to his definition of Emotion. Here, too, he has departed from the ordinary acceptation of the word; and, as we think, in the wrong direction. Whatever may be the interpretation that is justified by its derivation, the word Emotion has come generally to mean that kind of feeling which is not a direct result of any action on the organism; but is either an indirect result of such action, or arises quite apart from such action. It is used to indicate those sentient states which are independently generated in consciousness; as distinguished from those generated in our corporeal framework, and known as sensations. Now this distinction, tacitly made in common speech, is one which Psychology cannot well reject; but one which it must adopt, and to which it must give scientific precision. Mr. Bain, however, appears to ignore any such distinction. Under the term "emotion," he includes not only passions, sentiments, affections, but all "feelings, states of feeling, pleasures, pains," — that is, all sensations. This does not appear to be a mere lapse of expression; for when, in the opening sentence, he asserts that "mind is comprised under the three heads — Emotion, Volition, and Intellect," he of necessity implies that sensation is included under one of these heads; and as it cannot be included under Volition or Intellect, it must be classed with Emotion: as it clearly is in the next sentence.

We cannot but think tills is a retrograde step. Though distinctions which have been established in popular thought and language, are not unfrequently merged in the higher generalizations of science (as, for instance, when crabs and worms are grouped together in the sub-kingdom *Annulosa;)* yet science very generally recognizes the validity of these distinctions, as real though not fundamental. And so in the present case. Such community as analysis discloses between sensation and emotion, must not shut out the broad contrast that exists between them. If there needs a wider word, as there does, to signify any sentient state whatever; then we may fitly adopt for this purpose the word currently so used, namely, "Feeling." And considering as Feelings all that great division of mental states which we do not class as Cognitions, may then separate this great division into the two orders, Sensations and Emotions.

And here we may, before concluding, briefly indicate the leading outlines of a classification which reduces this distinction to a scientific form, and developes it somewhat further — a classification which, while suggested by certain fundamental traits reached without a very lengthened inquiry, is yet, we believe, in harmony with that disclosed by detailed analysis.

Leaving out of view the Will, which is a simple homogeneous mental state, forming the link between feeling and action, and not admitting of subdivisions; our states of consciousness fall into two great classes — Cognitions and Feelings.

Cognitions, or those modes of mind in which we are occupied with the *relations* that subsist among our feelings, are divisible into four great sub-classes.

Presentative cognitions; or those in which consciousness is occupied in localizing a sensation impressed on the organism — occupied, that is, with the relation between this presented mental state and those other presented mental states which make up our consciousness of the part affected: as when we cut ourselves.

Presentative-representative cognitions; or those in which consciousness is occupied with the relation between a sensation or group of sensations and the representations of those various other sensations that accompany it in experience. This is what we commonly call perception — an act in which, along with certain impressions presented to consciousness, there arise in consciousness the ideas of certain other impressions ordinarily connected with the presented ones: as when its visible form and colour, lead us to mentally endow an orange with all its other attributes.

Representative cognitions; or those in which consciousness is occupied with the relations among ideas or represented sensations: as in all acts of recollection.

Re-representative cognitions; or those in which the occupation of consciousness is not by representation of special relations, that have before been presented to consciousness; but those in which such represented special relations are thought of merely as comprehended in a general relation — those in which the concrete relations once experienced, in so far as they become objects of consciousness at all, are incidentally represented, along with the abstract relation which formulates them. The ideas resulting from this abstraction, do not themselves represent actual experiences; but are symbols which stand for groups of such actual experiences — represent aggregates of representations. And thus they may be called re-representative cognitions. It is clear that the process of re-representation is carried to higher stages, as the thought becomes more abstract.

Feelings, or those modes of mind in which we are occupied, not with the relations subsisting between our sentient states, but with the sentient states themselves, are divisible into four parallel sub-classes.

Presentative feelings, ordinarily called sensations, are those mental states in which, instead of regarding a corporeal impression as of this or that kind, or as located here or there, we contemplate it in itself as pleasure or pain: as when eating.

Presentative-representative feelings, embracing a great part of what we commonly call emotions, are those in which a sensation, or group of sensations or group of sensations and ideas, arouses a vast aggregation of represented sensations; partly of individual experience, but chiefly deeper than individual experience, and, consequently, indefinite. The emotion of terror may serve as an example. Along with certain impressions made on the eyes or ears, or both, are recalled in consciousness many of the pains to which such impressions have before been the antecedents and when the relation between such impressions and such pains has been habitual in the race, the definite ideas of inch pains which individual experience has given, are accompanied by the indefinite pains that result from inherited experience — vague feelings which we may call organic representations. In an infant, crying at a strange sight or sound while yet in the nurse's arms, we see these organic representations called into existence in the shape of dim discomfort, to which individual experience has yet given no specific outlines.

Representative feelings, comprehending the ideas of the feelings above classed, when they are called up apart from the appropriate external excitements. As instances of these may be named the feelings with which the descriptive poet writes, and which are aroused in the minds of his readers.

Re-representative feelings, under which head are included those more complex sentient states that are less the direct results of external excitements than the indirect or reflex results of them. The love of property is a feeling of this kind. It is awakened not by the presence of any special object, but by ownable objects at large; and it is not from the mere presence of such object, but from a certain ideal relation to them, that it arises. As before shown it consists, not of the represented advantages of possessing this or that, but of the represented advantages of possession in general — is not made up of certain concrete representations, but of the abstracts of many concrete representations; and so is re-representative. The higher sentiments, as that of justice, are still more completely of this nature. Here the sentient state is compounded out of sentient states that are themselves wholly, or almost wholly, re-representative: it involves representations of those lower emotions which are produced by the possession of property, by freedom of action, etc.; and thus is re-

representative in a higher degree.

This classification, here roughly indicated and capable of further expansion, will be found in harmony with the results of detailed analysis aided by development. Whether we trace mental progression through the grades of the animal kingdom, through the grades of mankind, or through the stages of individual growth; it is obvious that the advance, alike in cognitions and feelings, is, and must be, from the presentative to the more and more remotely representative. It is undeniable that intelligence ascends from those simple perceptions in which consciousness is occupied in localizing and classifying sensations, to perceptions more and more compound, to simple reasoning, to reasoning more and more complex and abstract — more and more remote from sensation. And in the evolution of feelings, there is a parallel series of steps. Simple sensations; sensations combined together; sensations combined with represented sensations; represented sensations organized into groups, in which their separate characters are very much merged; representations of these representative groups, in which the original components have become still more vague. In both cases, the progress has necessarily been from the simple and concrete to the complex and abstract: and as with the cognitions, so with the feelings, this must be the basis of classification.

The space here occupied with criticisms on Mr. Bain's work, we might have filled with exposition and eulogy, had we thought this the more important. Though we have freely pointed out what we conceive to be its defects, let it not be inferred that we question its great merits. We repeat that, as a natural history of the mind, we believe it to be the best yet produced. It is a most valuable collection of carefully-elaborated materials. Perhaps we cannot better express our sense of its worth, than by saying that, to those who hereafter give to this branch of Psychology a thoroughly scientific organization, Mr. Bain's book will be indispensable.

VIII
ILLOGICAL GEOLOGY

That proclivity to generalization which is possessed in greater or less degree by all minds, and without which indeed, intelligence cannot exist, has unavoidable inconveniences. Through it alone can truth be reached; and yet it almost inevitably betrays into error. But for the tendency to predicate of every other case, that which has been found in the observed cases, there could be no rational thinking; and yet by this indispensable tendency, men are perpetually led to found, on limited experience, propositions which they wrongly assume to be universal or absolute. In one sense, however, this can scarcely be regarded as an evil; for without premature generalizations the true generalization would never be arrived at. If we waited till all the facts were accumulated before trying to formulate them, the vast unorganized mass would be unmanageable. Only by provisional grouping can they be brought into such order as to be dealt with; and this provisional grouping is but another name for premature generalization.

How uniformly men follow this course, and how needful the errors are as steps to truth, is well illustrated in the history of Astronomy. The heavenly bodies move round the Earth in circles, said the earliest observers: led partly by the appearances, and partly by their experiences of central motions in terrestrial objects, with which, as all circular, they classed the celestial motions from lack of any alternative conception. Without this provisional belief, wrong as it was, there could not have been that comparison of positions which showed that the motions are not representable by circles; and which led to the hypothesis of epicycles and eccentrics. Only by the aid of this hypothesis, equally untrue, but capable of accounting more nearly for the appearances, and so of inducing more accurate observations — only thus did it become possible for Copernicus to show that the heliocentric theory is more feasible than the geocentric theory; or for Kepler to show that the planets move round the sun in ellipses. Yet again, without the aid of this approximate truth discovered by Kepler, Newton could not have established that general law from which it follows, that the motion of a heavenly body round its centre of gravity is not necessarily in an ellipse, but may be in any conic section. And lastly, it was only after the law of gravitation had been verified, that it became possible to determine the actual courses of planets, satellites, and comets; and to prove that, in consequence of perturbations, their orbits always deviate, more or less, from regular curves. Thus, there followed one another five provisional theories of the Solar System, before the sixth and absolutely true theory was reached. In which five provisional theories, each for a time held as final, we may trace both the tendency men have to leap from scanty data to wide generalizations, that are either untrue or but partially true; and the necessity there is for these transitional generalizations as steps to the final one.

In the progress of geological speculation the same laws of thought are clearly displayed. We have dogmas that were more than half false, passing current for a time as universal truths. We have evidence collected in proof of these dogmas; by and by a colligation of facts in antagonism with them; and eventually a consequent modification. In conformity with this somewhat improved hypothesis, we have a better classification of facts; a greater power of arranging and interpreting the new facts now rapidly gathered together; and further resulting corrections of hypothesis. Being, as we are at present, in the midst of this process, it is not possible to give an adequate account of the development of geological science as thus regarded: the earlier stages are alone known to us. Not only, however, is it interesting to observe how the more advanced views now received respecting the Earth's history, have been evolved out of the crude views which preceded them; but we shall find it extremely instructive to observe this. We shall see how greatly the old ideas still sway, both the general mind, and the minds of geologists themselves. We shall see how the kind of evidence that has in part abolished these old ideas, is still daily accumulating, and threatens to make other like revolutions. In brief, we shall see whereabouts we are in the elaboration of a true theory of the Earth; and, seeing our whereabouts, shall be the better able to judge, among various conflicting opinions, which best conform to the ascertained direction of geological discovery.

It is alike needless and impracticable here to enumerate the many speculations which were in earlier ages propounded by acute men — speculations some of which contained portions of truth. Falling in unfit times, these speculations did not germinate; and hence do not concern us. We have nothing to do with ideas, however good, out of which no science grew; but only with those which gave origin to the system of Geology that now exists. We therefore begin with Werner.

Taking for data the appearances of the Earth's crust in a narrow district of Germany; observing the constant order of superposition of strata, and their respective physical characters; Werner drew the inference that strata of like characters succeeded each other in like order over the entire surface of the Earth. And seeing, from the laminated structure of many formations and the organic remains contained in others, that they were sedimentary; he further inferred that these universal strata had been in succession precipitated from a chaotic menstruum which once covered our planet. Thus, on a very incomplete acquaintance with a thousandth part of the Earth's crust, he based a sweeping generalization applying to the whole of it. This Neptunist hypothesis, mark, borne out though it seemed to be by the most conspicuous surrounding facts, was quite untenable if analyzed. That a universal chaotic menstruum should deposit, one after another, numerous sharply-defined strata, differing from each other in composition, is incomprehensible. That the strata so deposited should contain the remains of plants and animals, which could not have lived under the supposed conditions, is still more incomprehensible. Physically absurd, however, as was this hypothesis, it recognized, though under a distorted form, one of the great agencies of geological change — that of water. It served also to express the fact that the formations of the Earth's crust stand in some kind of order. Further, it did a little towards supplying a nomenclature, without which much progress was impossible. Lastly, it furnished a standard with which successions of strata in various, regions could be compared, the differences noted, and the actual sections tabulated. It was the first provisional generalization; and was useful, if not indispensable, as a step to truer ones.

Following this rude conception, which ascribed geological phenomena to one agency, acting during one primeval epoch, there came a greatly-improved conception, which ascribed them to two agencies, acting alternately during successive epochs. Hutton, perceiving that sedimentary deposits were still being formed at the bottom of the sea from the detritus carried down by rivers; perceiving, further, that the strata of which the visible surface chiefly consists, bore marks of having been similarly formed out of pre-existing land; and inferring that these strata could have become land only by upheaval after their deposit; concluded that throughout an indefinite past, there had been periodic convulsions, by which continents were raised, with intervening eras of repose, during which such continents were worn down and transformed into new marine strata, fated to be in their turns elevated above the surface of the ocean. And finding that igneous action, to which sundry earlier geologists had ascribed basaltic rocks, was in countless places a source of disturbance, he taught that from it resulted these periodic convulsions. In this theory we see: — first, that the previously-recognized agency of water was conceived to act, not as by Werner, after a manner of which we have no experience, but after a manner daily displayed to us; and second, that the igneous agency, before considered only as a cause of special formations, was recognized as a universal agency, but assumed to act in an unproved way. Werner's sole process, Hutton developed from the catastrophic and inexplicable into the uniform and explicable; while that antagonistic second process, of which he first adequately estimated the importance, was regarded by him as a catastrophic one, and was not assimilated to known processes — not explained. We have here to note, however, that the facts collected and provisionally arranged in conformity with Werner's theory, served, after a time, to establish Hutton's more rational theory — in so far, at least, as aqueous formations, are concerned; while the doctrine of periodic subterranean convulsions, crudely as it was conceived by Hutton, was a temporary generalization needful as a step towards the theory of igneous action.

Since Hutton's time, the development of geological thought has gone still further in the same direction. These early sweeping doctrines have received additional qualifications. It has been discovered that more numerous and more heterogeneous agencies have been at work, than was at first believed. The igneous hypothesis has been rationalized, as the aqueous one had previously been: the gratuitous assumption of vast elevations suddenly occurring after long intervals of quiescence, has grown into the consistent theory, that islands and continents are the accumulated results of successive small upheavals, like those experienced in ordinary earthquakes.

To speak more specifically, we find, in the first place, that instead of assuming the denudation produced by rain and rivers to be the sole means of wearing down lands and producing their irregularities of surface, geologists now see that denudation is only a part-cause of such irregularities; and further, that the new strata deposited at the bottom of the sea, are not the products of river-sediment solely, but are in part due to the action of waves and tidal currents on the coasts. In the second place, we find that Hutton's conception of upheaval by subterranean forces, has not only been modified by assimilating these subterranean forces to ordinary earthquake-forces; but modern inquiries have shown that, besides elevations of surface, subsidences are thus produced; that local upheavals, as well as the general upheavals, which raise continents, come within the same category; and that all these changes are probably consequent on the progressive collapse of the Earth's crust upon its cooling and contracting nucleus — the only adequate cause. In the third place, we find that beyond these two great antagonist agencies, modern Geology recognises sundry minor ones: as those of glaciers and icebergs; those of coral-polypes; those of *Protozoa* having siliceous or calcareous shells — each of which agencies, insignificant as it seems, is found capable of slowly working terrestrial changes of considerable magnitude. Thus, then, the recent progress of Geology has been a still further departure from primitive conceptions. Instead of one catastrophic cause, once in universal action, as supposed by Werner — instead of one general continuous cause, antagonized at long intervals by a catastrophic cause, as taught by Hutton; we now recognize several causes, all more or less general and continuous. We no longer resort to hypothetical agencies to explain the phenomena displayed by the Earth's crust; but we are day by day more clearly perceiving that these phenomena have arisen from forces like those now at work, which have acted in all varieties of combination, through immeasurable periods of time.

Having thus briefly traced the evolution of geologic science, and noted its present form, let us go on to observe the way in which it is still swayed by the crude hypotheses it set out with; so that even now, old doctrines that are abandoned as untenable in theory, continue in practice to mould the ideas of geologists, and to foster sundry beliefs that are logically indefensible. We shall see, both how those simple sweeping conceptions with which the science commenced, are those which every student is apt at first to seize hold of, and how several influences conspire to maintain the twist thus resulting — how the original nomenclature of periods and formations necessarily keeps alive the original implications; and how the need for arranging new data in some order, naturally results in their being thrust into the old classification, unless their incongruity with it is very glaring. A few facts will best prepare the way for criticism.

Up to 1839 it was inferred, from their crystalline character, that the metamorphic rocks of Anglesea are more ancient than any rocks of the adjacent main land; but it has since been shown that they are of the same age with the slates and grits of Carnarvon and Merioneth. Again, slaty cleavage having been first found only in the lowest rocks, was taken as an indication of the highest antiquity: whence resulted serious mistakes; for this mineral characteristic is now known to occur in the Carboniferous system. Once more, certain red conglomerates and grits on the north-west coast of Scotland, long supposed from their lithological aspect to belong to the Old Red Sandstone, are now identified with the Lower Silurians.

These are a few instances of the small trust to be placed in mineral qualities, as evidence of the ages or relative positions of strata. From the recently-published third edition of *Siluria,* may be culled numerous facts of like implication. Sir R. Murchison considers it ascertained, that the siliceous Stiper stones of Shropshire are the equivalents of the Tremadock slates of North Wales. Judged by their fossils, Bala slate and limestone are of the same age as the Caradoc sandstone, lying forty miles off. In Radnorshire, the formation classed as upper Llandovery rock, is described at different spots, as "sandstone or conglomerate," "impure limestone," "hard coarse grits," "siliceous grit" — a considerable variation for so small an area as that of a county. Certain sandy beds on the left bank of the Towy, which Sir R. Murchison had, in his *Silurian System,* classed as Caradoc sandstone (evidently from their mineral characters), he now finds, from their fossils, belong to the Llandeilo formation. Nevertheless, inferences from mineral characters are still habitually drawn and received. Though *Siluria,* in common with other geological works, supplies numerous proofs that rocks of the same age are often of widely-different composition a few miles off, while rocks of widely different ages are often of similar composition; and though Sir. R. Murchison shows us, as in the case just cited, that he has himself in past times been misled by trusting to lithological evidence; yet his reasoning, all through *Siluria,* shows that he still thinks it natural to expect formations of the same age to be chemically similar, even in remote regions. For example, in treating of the Silurian rocks of South Scotland, he says: — "When traversing the tract between Dumfries and Moffat in 1850, it occurred to me that the dull reddish or purple sandstone and schist to the north of the former town, which so resembled the bottom rocks of the Longmynd, Llanberis, and St. David's, would prove to be of the same age;" and further on ho again insists upon the fact that these strata "are absolutely of the same composition as the bottom rocks of the Silurian region."

On this unity of mineral character it is, that this Scottish formation is concluded to be contemporaneous with the lowest formations in Wales; for the scanty palæontological evidence suffices neither for proof nor disproof. Now, had there been a continuity of like strata in like order between Wales and Scotland, there might have been little to criticise in this conclusion. But since Sir R. Murchison himself admits, that in Westmoreland and Cumberland, some members of the system "assume a lithological aspect different from what they maintain in the Silurian and Welsh region," there seems no reason to expect mineralogical continuity in Scotland. Obviously therefore, the assumption that these Scottish formations are of the same age with the Longmynd of Shropshire, implies the latent belief that certain mineral characters indicate certain eras.

Far more striking instances, however, of the influence of this latent belief remain to be given. Not in such comparatively near districts as the Scottish lowlands only, does Sir R. Murchison expect a repetition of the Longmynd strata; but in the Rhenish provinces, certain "quartzose flagstones and grits, like those of the Longmynd," are seemingly concluded to be of contemporaneous origin, because of their likeness. "Quartzites in roofing-slates with a greenish tinge that reminded us of the lower slates of Cumberland and Westmoreland," are evidently suspected to be of the same age. In Russia, he remarks that the carboniferous limestones "are overlaid along the western edge of the Ural chain by sandstones and grits, which occupy much the same place in the general series as the millstone grit of England;" and in calling this group, as he does, the "representative of the millstone grit," Sir R. Murchison clearly shows that he thinks likeness of mineral composition some evidence of equivalence in time, even at that great distance. Nay, on the flanks of the Andes and in the United States, such similarities are looked for, and considered as significant of certain ages. Not that Sir R. Murchison contends theoretically for this relation between lithological character and date. For on the page from which we have just quoted *(Siluria)*, he says, that "whilst the soft Lower Silurian clays and sands of St. Petersburg have their equivalents in the hard schists and quartz rocks with gold veins in the heart of the Ural mountains, the equally soft red and green Devonian marls of the Valdai Hills are represented on the western flank of that chain, by hard, contorted, and fractured limestones." But these, and other such admissions, seem to go for little. Whilst himself asserting that the Potsdam-sandstone of North America, the Lingula-flags of England, and the alum-slates of Scandinavia are of the same period — while fully aware that among the Silurian formations of Wales, there are oolitic strata like those of secondary age; yet is his reasoning more or less coloured by the assumption, that formations of like qualities probably belong to the same era. Is it not manifest, then, that the exploded hypothesis of Werner continues to influence geological speculation?

"But," it will perhaps be said, "though individual strata are not continuous over large areas, yet systems of strata are. Though within a few miles the same bed gradually passes from clay into sand, or thins out and disappears, yet the group of strata to which it belongs does not do so; but maintains in remote regions the same relations to other groups."

This is the generally-current belief. On this assumption the received geological classifications appear to be framed. The Silurian system, the Devonian system, the Carboniferous system, etc., are set down in our books as groups of formations which everywhere succeed each other in a given order; and are severally everywhere of the same age. Though it may not be asserted that these successive systems are universal; yet it seems to be tacitly assumed that they are so. In North and South America, in Asia, in Australia, sets of strata are assimilated to one or other of these groups; and their possession of certain mineral characters and a certain order of superposition are among the reasons assigned for so assimilating them. Though, probably, no competent geologist would contend that the European classification of strata is applicable to the globe as a whole; yet most, if not all geologists, write as though it were so. Among readers of works on Geology, nine out ten carry away the impression that the divisions, Primary, Secondary and Tertiary, are of absolute and uniform application; that these great divisions are separable into subdivisions, each of which is definitely distinguishable from the rest, and is everywhere recognizable by its characters as such or such; and that in all parts of the Earth, these minor systems severally began and ended at the same time. When they meet with the term "carboniferous era," they take for granted that it was an era universally carboniferous — that it was, what Hugh Miller indeed actually describes it, an era when the Earth bore a vegetation far more luxuriant than it has since done; and were they in any of our colonies to meet with a coal-bed, they would conclude at, as a matter of course, it was of the same age as the English coal-beds.

Now this belief that geologic "systems" are universal, is quite as untenable as the other. It is just as absurd when considered *à priori;* and it is equally inconsistent with the facts. Though some series of strata classed together as Oolite, may range over a wider district than any one stratum of the series; yet we have but to ask what were the circumstances of its deposit, to see that the Oolitic series, like one of its individual strata, must be of local origin; and that there is not likely to be anywhere else, a series that exactly corresponds, either in its characters or in its commencement and termination. For the formation of such a series implies an area of subsidence, in which its component beds were thrown down. Every area of subsidence is necessarily limited; and to suppose that there exist elsewhere groups of beds completely answering to those known as Oolite, is to suppose that, in contemporaneous areas of subsidence, like processes were going on. There is no reason to suppose this; but every reason to suppose the reverse. That in contemporaneous areas of subsidence throughout the globe, the conditions would cause the formation of Oolite, or anything like it, is an assumption which no modern geologist would openly make: he would say that the equivalent series of beds found elsewhere, would very likely be of dissimilar mineral character.

Moreover, in these contemporaneous areas of subsidence, the phenomena going on would not only be more or less different in kind; but in no two cases would they be likely to agree in their commencements and terminations. The probabilities are greatly against separate portions of the Earth's surface beginning to subside at the same time, and ceasing to subside at the same time — a coincidence which alone could produce equivalent groups of strata. Subsidences in different places begin and end with utter irregularity; and hence the groups of strata thrown down in them can but rarely correspond. Measured against each other in time, their limits will disagree. They will refuse to fit into any scheme of definite divisions. On turning to the evidence, we find that it daily tends more and more to justify these *à priori* positions. Take, as an example, the Old Red Sandstone system. In the north of England this is represented by a single stratum of conglomerate. In Herefordshire, Worcestershire, and Shropshire, it expands into a series of strata from eight to ten thousand feet thick, made up of conglomerates, red, green, and white sandstones, red, green, and spotted, marls, and concretionary limestones. To the south-west, as between Caermarthen and Pembroke, these Old Red Sandstone strata exhibit considerable lithological changes; and there is an absence of fossil fishes. On the other side of the Bristol Channel, they display further changes in mineral characters and remains. While in South Devon and Cornwall, the equivalent strata, consisting chiefly of slates, schists, and limestones, are so wholly different, that they were for a long time classed as Silurian. When we thus see that in certain directions the whole group of deposits thins out, and that its mineral characters as well as its fossils change within moderate distances; does it not become clear that the whole group of deposits was a local one? And when we find, in other regions, formations analogous to these Old Red Sandstone or

Devonian formations; is it certain — is it even probable — that they severally began and ended at the same time with them? Should it not require overwhelming evidence to make us believe as much?

Yet so strongly is geological speculation swayed by the tendency to regard the phenomena as general instead of local, that even those most on their guard against it seem unable to escape its influence. At page 158 of his *Principles of Geology,* Sir Charles Lyell says: —

"A group of red marl and red sandstone, containing salt and gypsum, being interposed in England between the Lias and the Coal, all other red marls and sandstones, associated some of them with salt, and others with gypsum, and occurring not only in different parts of Europe, but in North America, Peru, India, the salt deserts of Asia, those of Africa — in a word, in every quarter of the globe, were referred to one and the same period....It was in vain to urge as an objection the improbability of the hypothesis which implies that all the moving waters on the globe were once simultaneously charged with sediment of a red colour. But the rashness of pretending to identify, in age, all the red sandstones and marls in question, has at length been sufficiently exposed, by the discovery that, even in Europe, they belong decidedly to many different epochs."

Nevertheless, while in this and numerous passages of like implication. Sir C. Lyell protests against the bias here illustrated, he seems himself not completely free from it. Though he utterly rejects the old hypothesis that all over the Earth the same continuous strata lie upon each other in regular order, like the coats of an onion, he still writes as though geologic "systems" do thus succeed each other. A reader of his *Manual* would certainly suppose him to believe, that the Primary epoch ended, and the Secondary epoch commenced, all over the world at the same time — that these terms really correspond to distinct universal eras in Nature. When he assumes, as he does, that the division between Cambrian and Lower Silurian in America, answers chronologically to the division between Cambrian and Lower Silurian in Wales — when he takes for granted that the partings of Lower from Middle Silurian, and of Middle Silurian from Upper, in the one region, are of the same dates as the like partings in the other region; does it not seem that he believes geologic "systems" to be universal, in the sense that their separations were in all places contemporaneous? Though he would, doubtless, disown this as an article of faith, is not his thinking unconsciously influenced by it? Must we not say that though the onion-coat hypothesis is dead, its spirit is traceable, under a transcendental form, even in the conclusions of its antagonists?

Let us now consider another leading geological doctrine, introduced to us by the cases just mentioned. We mean the doctrine that strata of the same age contain like fossils; and that, therefore, the age and relative position of any stratum may be known by its fossils. While the theory that strata of like mineral characters were everywhere deposited simultaneously, has been ostensibly abandoned, there has been accepted the theory that in each geologic epoch similar plants and animals existed everywhere; and that, therefore, the epoch to which any formation belongs may be known by the organic remains contained in the formation. Though, perhaps, no leading geologist would openly commit himself to an unqualified assertion of this theory, yet it is tacitly assumed in current geological reasoning.

This theory, however, is scarcely more tenable than the other. It cannot be concluded with any certainty, that formations in which similar organic remains are found, were of contemporaneous origin; nor can it be safely concluded that strata containing different organic remains are of different ages. To most readers these will be startling propositions; but they are fully admitted by the highest authorities. Sir Charles Lyell confesses that the test of organic remains must be used "under very much the same restrictions as the test of mineral composition." Sir Henry de la Beche, who variously illustrates this truth, gives, as one instance, the great incongruity there must be between the fossils of our carboniferous rooks and those of the marine strata deposited at the same period. But though, in the abstract, the danger of basing positive conclusions on evidence derived from fossils, is clearly recognized; yet, in the concrete, this danger is generally disregarded. The established conclusions respecting the ages of strata, take but little note of it; and by some geologists it seems altogether ignored. Throughout his *Siluria,* Sir R. Murchison habitually assumes that the same, or kindred, species, lived in all parts of the Earth at the same time. In Russia, in Bohemia, in the United States, in South America, strata are classed as belonging to this or that part of the Silurian system, because of the similar fossils contained in them — are concluded to be everywhere contemporaneous if they enclose a proportion of identical or allied forms. In Russia the relative position of a stratum is inferred from the fact that, along with some Wenlock forms, it yields the *Pentamerus oblongus.* Certain crustaceans called Eurypteri, being characteristic of the Upper Ludlow rock, it is remarked that "large Eurypteri occur in a so-called black grey-wacke slate in Westmoreland, in Oneida County, New York, which will probably be found to be on the parallel of the Upper Ludlow rock:" in which word "probably," we see both how dominant is this belief of universal distribution of similar creatures at the same period, and how apt this belief is to make its own proof, by raising the expectation that the ages are identical when the forms are alike. Besides thus interpreting the formations of Russia, England, and America, Sir R. Murchison thus interprets those of the antipodes. Fossils from Victoria Colony, he agrees with the Government-surveyor in classing as of Lower Silurian or Llandovery age: that is, he takes for granted that when certain crustaceans and mollusks were living in Wales, certain similar crustaceans and mollusks were living in Australia.

Yet the improbability of this assumption may be readily shown from Sir R. Murchison's own facts. If, as he points out, the crustacean fossils of the uppermost Silurian rocks in Lanarkshire are, "with one doubtful exception," "all distinct from any of the forms on the same horizon in England;" how can it be fairly presumed that the forms existing on the other side of the Earth during the Silurian period, were nearly allied to those existing here? Not only, indeed, do Sir R. Murchison's conclusions tacitly assume this doctrine of universal distribution, but he distinctly enunciates it. "The mere presence of a graptolite," he says, "will at once decide that the enclosing rock is Silurian;" and he says this, notwithstanding repeated warnings against such generalizations. During the progress of Geology, it has over and over again happened that a particular fossil, long considered characteristic of a particular formation, has been afterwards discovered in other formations. Until some twelve years ago, Goniatites had not been found lower than the Devonian rocks; but now, in Bohemia, they have been found in rocks classed as Silurian. Quite recently, the Orthoceras, previously supposed to be a type exclusively palæozoic, has been detected along with mesozoic Ammonites and Belemnites. Yet hosts of such experiences fail to extinguish the assumption, that the age of a stratum may be determined by the occurrence in it of a single fossil form.

Nay, this assumption survives evidence of even a still more destructive kind. Referring to the Silurian system in Western Ireland, Sir R. Murchison says, "in the beds near Maam, Professor Nicol and myself collected remains, some of which would be considered Lower, and others Upper, Silurian;" and he then names sundry fossils which, in England, belong to the summit of the Ludlow rocks, or highest Silurian strata; "some, which elsewhere are known only in rocks of Llandovery age," that is, of middle Silurian age; and some, only before known in Lower Silurian strata, not far above the most ancient fossiliferous beds. Now what do these facts prove? Clearly, they prove that species which in Wales are separated by strata more than twenty thousand feet deep, and therefore seem to belong to periods far more remote from each other, were really coexistent. They prove that the mollusks and crinoids held characteristic of early Silurian strata, and supposed to have become extinct long before the mollusks and crinoids of the later Silurian strata came into existence, were really flourishing at the same time with these last; and that these last possibly date back to as early a period as the first. They prove that not only the mineral characters of sedimentary formations, but also the collections of organic forms they contain, depend, to a great extent, on local circumstances. They prove that the fossils met with in any series of strata, cannot be taken as representing anything like the whole Flora and Fauna of the period they belong to. In brief, they throw great doubt upon numerous geological generalizations.

Notwithstanding facts like these, and notwithstanding his avowed opinion that the test of organic remains must be used "under very much the same restrictions as the test of mineral composition," Sir Charles Lyell, too, bases positive conclusions on this test: even where the community of fossils is slight and the distance great. Having decided that in various places in Europe, middle Eocene strata are distinguished by nummulites; he infers, without any other assigned evidence, that wherever nummulites are found — in Morocco, Algeria, Egypt, in Persia, Scinde, Cutch, Eastern Bengal, and the frontiers of China — the containing formation is middle Eocene. And from this inference he draws the following important corollary: —

"When we have once arrived at the conviction that the nummulitic formation occupies a middle place in the Eocene series, we are struck with the comparatively modern date to which some of the greatest revolutions in the physical geography of Europe, Asia, and northern Africa must be referred. All the mountain chains, such as the Alps, Pyrenees, Carpathians, and Himalayas, into the composition of whose central and loftiest parts the nummulitic strata enter bodily, could have had no existence till after the middle Eocene period." — *Manual,* p. 232.

A still more marked case follows on the next page. Because a certain bed at Claiborne in Alabama, which contains "*four hundred* species of marine shells," includes among them the *Cardita planicosta,*" and *some others* identical with European species, or very nearly allied to them," Sir C. Lyell says it is "highly probable the Claiborne beds agree in age with the central or Bracklesham group of England." When we find contemporaneity supposed on the strength of a community no greater than that which sometimes exists between strata of widely-different ages in the same country, it seems very much as though the above-quoted caution had been forgotten. It appears to be assumed for the occasion, that species which had a wide range in space had a narrow range in time; which is the reverse of the fact. The tendency to systematize overrides the evidence, and thrusts Nature into a formula too rigid to fit her endless variety.

"But," it may be urged, "surely, when in different places the order of superposition, the mineral characters, and the fossils, agree, it may be safely concluded that the formations thus corresponding are equivalents in time. If, for example, the United States display the same succession of Silurian, Devonian, and Carboniferous systems, lithologically similar, and characterized by like fossils, it is a fair inference that these groups of strata were severally deposited in America at the same periods that they were deposited here."

On this position, which seems a strong one, we have, in the first place, to remark, that the evidence of correspondence is always more or less suspicious. We have already adverted to the several "idols" — if we may use Bacon's metaphor — to which geologists unconsciously sacrifice, when interpreting the structures of unexplored regions. Carrying with them the classification of strata existing in Europe, and assuming that groups of strata in other parts of the world must answer to some of the groups of strata known here, they are necessarily prone to assert parallelism on insufficient evidence. They scarcely entertain the previous question, whether the formations they are examining have or have not any European equivalents; but the question is — with which of the European series shall they be classed? — with which do they most agree? — from which do they differ least? And this being the mode of enquiry, there is apt to result great laxity of interpretation. How lax the interpretation really is, may be readily shown. When strata are discontinuous, as between Europe and America, no evidence can be derived from the order of superposition, apart from mineral characters and organic remains; for, unless strata can be continuously traced, mineral characters and organic remains are the only means of classing them as such or such.

As to the test of mineral characters, we have seen that it is almost worthless; and no modern geologist would dare to say it should be relied on. If the Old Red Sandstone series in mid-England, differs wholly in lithological aspect from the equivalent series in South Devon, it is clear that similarities of texture and composition can have no weight in assimilating a system of strata in another quarter of the globe to some European system. The test of fossils, therefore, is the only one that remains; and with how little strictness this test is applied, one case will show. Of forty-six species of British Devonian corals, only six occur in America; and this, notwithstanding the wide range which the Anthozoa are known to have. Similarly of the Mollusca and Crincidea, it appears that, while then are sundry genera found in America that are found here, there are scarcely any of the same species. And Sir Charles Lyell admits that "the difficulty of deciding on the exact parallelism of the New York subdivisions, as above enumerated, with the members of the European Devonian, is very great, so few are the species in common." Yet it is on the strength of community of fossils, that the whole Devonian series of the United States is assumed to be contemporaneous with the whole Devonian series of England. And it is partly on the ground that the Devonian of the United. States corresponds in time with our Devonian, that Sir Charles Lyell concludes the superjacent coal-measures of the two countries to be of the same age. Is it not, then, as we said, that the evidence in these cases is very suspicious?

Should it be replied, as it may fairly be, that this correspondence from which the synchronism of distant formations is inferred, is not a correspondence between particular species or particular genera, but between the general characters of the contained assemblages of fossils — between the *facies* of the two Faunas; the rejoinder is, that though such correspondence is a stronger evidence of synchronism it is still an insufficient one. To infer synchronism from such correspondence, involves the postulate that throughout each geologic era there has habitually existed a recognizable similarity between the groups of organic forms inhabiting all the different parts of the Earth; and that the causes which have in one part of the Earth changed the organic forms into those which characterize the next era, have simultaneously acted in all other parts of the Earth, in such ways as to produce parallel changes of their organic forms. Now this is not only a large assumption to make; but it is an assumption contrary to probability. The probability is, that the causes which have changed Faunas have been local rather than universal; that hence while the Faunas of some regions have been rapidly changing, those of others have been almost quiescent; and that when such others have been changed, it has been, not in such ways as to maintain parallelism, but in such ways as to produce divergence.

Even supposing, however, that districts some hundreds of miles apart, furnished groups of strata that completely agreed in their order of superposition, their mineral characters, and their fossils, we should still have inadequate proof of contemporaneity. For there are conditions, very likely to occur, under which such groups might differ widely in age. If there be a continent of which the strata crop out on the surface obliquely to the line of coast — running, say, west-northwest, while the coast runs east and west — it is clear that each group of strata will crop out on the beach at a particular part of the coast; that further west the next group of strata will crop out on the beach; and so continuously. As the localization of marine plants and animals is in a considerable degree determined by the nature of the rocks and their detritus, it follows that each part of this coast will have its more or less distinct Flora and Fauna. What now must result from the action of the waves in the course of a geologic epoch? As the sea makes slow inroads on the land, the place at which each group of strata crops out on the beach will gradually move towards the west: its distinctive fish, mollusks, crustaceans, and sea-weeds, migrating with it. Further, the detritus of each of these groups of strata will, as the point of outcrop moves westwards, be deposited over the detritus of the group in advance of it. And the consequence of these actions, carried on for one of those enormous periods required for geologic changes, will be that, corresponding to each eastern stratum, there will arise a stratum far to the west which, though occupying the same position relatively to other beds, formed of like materials, and containing like fossils, will yet be perhaps a million years later in date.

But the illegitimacy, or at any rate the great doubtfulness, of many current geological inferences, is best seen when we contemplate terrestrial changes now going on: and ask how far such inferences are countenanced by them. If we carry out rigorously the modern method of interpreting geological phenomena, which Sir Charles Lyell has done so much to establish — that of referring them to causes like those at present in action — we cannot fail to see how improbable are sundry of the received conclusions.

Along each line of shore that is being worn away by the waves, there are being formed mud, sand, and pebbles. This detritus, spread over the neighbouring sea-bottom, has, in each locality, a more or less special character; determined by the nature of the strata destroyed. In the English Channel it is not the same as in the Irish Channel; on the east coast of Ireland it is not the same as on the west coast; and so throughout. At the mouth of each great river, there is being deposited sediment differing more or less from that of other rivers in colour and quality; forming strata that are here red, there yellow, and elsewhere brown, grey, or dirty white. Besides which various formations, going on in deltas and along shores, there are some much wider and still more contrasted formations. At the bottom of the Ægæan Sea, there is accumulating a bed of Pteropod shells, which will eventually, no doubt, become a calcareous rock. For some hundreds of thousands of square miles, the ocean-bed between Great Britain and North America, is being covered with a stratum of chalk;

and over large areas in the Pacific, there are going an deposits of coralline limestone. Thus, throughout the Earth, there are at this moment being prod deed an immense number of strata differing from each other in lithological characters. Name at random any one part of the sea-bottom, and ask whether the deposit there taking place is like the deposit taking place at some distant part of the sea-bottom, and the almost-certainly correct answer will be — No. The chances are not in favour of similarity, but very greatly against it.

In the order of superposition of strata there is occuring a like variety. Each region of the Earth's surface has its special history of elevations, subsidences, periods of rest; and this history in no case fits chronologically with the history of any other portion. River deltas are now being thrown down on formations of quite different ages. While here there has been deposited a series of beds many hundreds of feet thick, there has elsewhere been deposited but a single bed of fine mud. While one region of the Earth's crust, continuing for a vast epoch above the surface of the ocean, bears record of no changes save those resulting from denudation; another region of the Earth's crust gives proof of various changes of level, with their several resulting masses of stratified detritus. If anything is to be judged from current processes, we must infer, not only that everywhere the succession of sedimentary formations differs more or less from the succession elsewhere; but also that in each place, there exist groups of strata to which many other places have no equivalents.

With respect to the organic bodies imbedded in formations now in progress, the like truth is equally manifest, if not more manifest. Even along the same coast, within moderate distances, the forms of life differ very considerably; much more on coasts that are remote from each other. Again, dissimilar creatures that are living together near the same shore, do not leave their remains in the same beds of sediment. For instance, at the bottom of the Adriatic, where the prevailing currents cause the deposits to be here of mud, and there of calcareous matter, it is proved that different species of co-existing shells are being buried in these respective formations. On our own coasts, the marine remains found a few miles from shore, in banks where fish congregate, are different from those found close to the shore, where only littoral species flourish. A large proportion of aquatic creatures have structures that do not admit of fossilization; while of the rest, the great majority are destroyed, when dead, by the various kinds of scavengers that creep among the rocks and weeds. So that no one deposit near our shores can contain anything like a true representation of the Fauna of the surrounding sea; much less of the co-existing Faunas of other seas in the same latitude; and still less of the Faunas of seas in distant latitudes. Were it not that the assertion seems needful, it would be almost absurd to say, that the organic remains now being buried in the Dogger Bank, can tell us next to nothing about the fish, crustaceans, mollusks, and corals that are being buried in the Bay of Bengal.

Still stronger is the argument in the case of terrestrial life. With more numerous and greater contrasts between the plants and animals of remote places, there is a far more imperfect registry of them. Schouw marks out on the Earth more than twenty botanical regions, occupied by groups of forms so far distinct from each other, that, if fossilized, geologists would scarcely be disposed to refer them all to the same period. Of Faunas, the Arctic differs from the Temperate; the Temperate from the Tropical; and the South Temperate from the North Temperate. Nay, in the South Temperate Zone itself, the two regions of South Africa and South America are unlike in their mammals, birds, reptiles, fishes, mollusks, insects. The shells and bones now lying at the bottoms of lakes and estuaries in these several regions have certainly not that similarity which is usually looked for in those of contemporaneous strata; and the recent forms exhumed in any one of these regions would very untruly represent the present Flora and Fauna of the Earth. In conformity with the current style of geological reasoning, an exhaustive examination of deposits in the Arctic circle, might be held to prove that though at this period there were sundry mammals existing, there were no reptiles; while the absence of mammals in the deposits of the Galapagos Archipelago, where there are plenty of reptiles, might be held to prove the reverse. And at the same time, from the formations extending for two thousand miles along the great barrier-reef of Australia — formations in which are imbedded nothing but corals, echinoderms, mollusks, crustaceans, and fish, along with an occasional turtle, or bird, or cetacean, it might be inferred that there lived in our epoch neither terrestrial reptiles nor terrestrial mammals.

The mention of Australia, indeed, suggests an illustration which, even alone, would amply prove our case. The Fauna of this region differs widely from any that is found elsewhere. On land all the indigenous mammals, except bats, belong to the lowest, or implacental division; and the insects are singularly different from those found elsewhere. The surrounding seas contain numerous forms that are more or less strange; and among the fish there exists a species of shark, which is the only living representative of a genus that flourished in early geologic epochs. If, now, the modern fossiliferous deposits of Australia were to be examined by one ignorant of the existing Australian Fauna; and if he were to reason in the usual manner; he would be very unlikely to class these deposits with those of the present time. How, then, can we place confidence in the tacit assumption that certain formations in remote parts of the Earth are referable to the same period, because the organic remains contained in them display a certain community of character? or that certain others are referable to different periods, because the *facies* of their Faunas are different?

"But," it will be replied, "in past eras the same, or similar, organic forms were more widely distributed than now." It may be so; but the evidence adduced by no means proves it. The argument by which this conclusion is reached, runs a risk of being quoted as an example of reasoning in a circle. As already pointed out, between formations in remote regions there is no means of ascertaining equivalence but by fossils. If, then, the contemporaneity of remote formations is concluded from the likeness of their fossils; how can it be said that similar plants and animals were once more widely distributed, because they are found in contemporaneous strata in remote regions? Is not the fallacy manifest? Even supposing there were no such fatal objection as this, the evidence commonly assigned would still be insufficient. For we must bear in mind that the community of organic remains commonly thought sufficient for inferring correspondence in time, is a very imperfect community. When the compared sedimentary beds are far apart, it is scarcely expected that there will be many species common to the two: it is enough if there be discovered a considerable number of common genera. Now had it been proved that, throughout geologic time, each genus lived but for a short period — a period measured by a single group of strata — something might be inferred. But what if we learn that many of the same genera continued to exist throughout enormous epochs, measured by several vast systems of strata? "Among molluscs, the genera *Avicula, Modiola, Terebratula, Lingula,* and *Orbicula,* are found from the Silurian rocks upwards to the present day." If, then, between the lowest fossiliferous formations and the most recent, there exists this degree of community; must we not infer that there will probably often exist a degree of community between strata that are far from contemporaneous?

Thus the reasoning from which it is concluded that similar organic forms were once more widely spread, is doubly fallacious; and, consequently, the classifications of foreign strata based on this conclusion are untrustworthy. Judging from the present distribution of life, we can scarcely expect to find similar remains in geographically remote strata of the same age; and where, between the fossils of geographically remote strata, we do find much similarity, it is probably often due rather to likeness of conditions than to contemporaneity. If from causes and effects such as we now witness, we reason back to the causes and effects of past epochs, we discover inadequate warrant for sundry of the received doctrines. Seeing, as we do, that in large areas of the Pacific this is a period characterized by abundance of corals; that in the North Atlantic it is a period in which a great chalk-deposit is being formed; and that in the valley of the Mississippi it is a period of new coal-basins — seeing also, as we do, that in one extensive continent this is peculiarly an era of implacental mammals, and that in another extensive continent it is peculiarly an era of placental mammals; we have good reason to hesitate before accepting these sweeping generalizations which are based on a cursory examination of strata occupying but a tenth part of the Earth's surface.

At the outset, this article was to have been a review of the works of Hugh Miller; but it has grown into something much more general. Nevertheless, the remaining two doctrines which we propose to criticise, may be conveniently treated, in connection with his name, as that of one who fully committed himself to them. And first, a few words with regard to his position.

That he was a man whose life was one of meritorious

achievement, every one knows. That he was a diligent and successful working geologist, scarcely needs saying. That with indomitable perseverance he struggled up from obscurity to a place in the world of literature and science, shows him to have been highly endowed in character and intelligence. And that he had a remarkable power of presenting his facts and arguments in an attractive form, a glance at any of his books will quickly prove. By all means, let us respect him as a man of activity and sagacity, joined with a large amount of poetry. But while saying this we must add, that his reputation stands by no means so high in the scientific world as in the world at large. Partly from the fact that our Scotch neighbours are in the habit of blowing the trumpet rather loudly before their notabilities — partly because the charming style in which his books are written has gained him a large circle of readers — partly, perhaps, through a praiseworthy sympathy with him as a self-made man; Hugh Miller has met with an amount of applause which, little as we wish to diminish it, must not be allowed to blind the public to his defects as a man of science.

The truth is, he was so far committed to a foregone conclusion, that he could not become a philosophical geologist. He might be aptly described as a theologian studying geology. The dominant idea with which he wrote, may be seen in the titles of his books — *Law versus Miracle, — Footprints of the Creator, — The Testimony of the Rocks.* Regarding geological facts as evidence for or against certain religious conclusions, it was scarcely possible for him to deal with geological facts impartially. His ruling aim was to disprove the Development Hypothesis, the assumed implications of which were repugnant to him; and in proportion to the strength of his feeling, was the one-sidedness of his reasoning. He admitted that "God might as certainly have *originated* the species by a law of development, as he *maintains* it by a law of development; the existence of a First Great Cause is as perfectly compatible with the one scheme as with the other." Nevertheless, he considered the hypothesis at variance with Christianity; and therefore combated with it. He apparently overlooked the fact that the doctrines of geology in general, as held by himself, had been rejected by many on similar grounds; and that he had himself been repeatedly attacked for his anti-Christian teachings. He seems not to have perceived that, just as his antagonists were wrong in condemning as irreligious, theories which he saw were not irreligious; so might he be wrong in condemning, on like grounds, the Theory of Evolution. In brief, he fell short of that highest faith, which knows that all truths must harmonize; and which is, therefore, content trustfully to follow the evidence whithersoever it leads.

Of course it is impossible to criticize his works without entering on this great question to which he chiefly devoted himself. The two remaining doctrines to be here discussed, bear directly on this question; and, as above said, we propose to treat them in connection with Hugh Miller's name, because, throughout his reasonings, he assumes their truth. Let it not be supposed, however, that we shall aim to prove what he has aimed to disprove. While we purpose showing that his arguments against the Development Hypothesis are based on invalid assumptions; we do not purpose showing that the opposing arguments are based on valid assumptions. We hope to make it apparent that the geological evidence at present obtained, is insufficient for either side; further, that there seems little probability of sufficient evidence ever being obtained; and that if the question is eventually decided, it must be decided on other than geological data.

The first of the current doctrines to which we have just referred, is, that there occur in the records of former life on our planet, certain great blanks — that though, generally, the succession of fossil forms is tolerably continuous, yet that at two places there occur wide gaps in the series whence it is inferred that, on at least two occasions, the previously existing inhabitants of the Earth were almost wholly destroyed, and a different class of inhabitants created. Comparing the general life on the Earth to a thread, Hugh Miller says: —

"It is continuous from the present time up to the commencement of the Tertiary period; and then so abrupt a break occurs, that, with the exception of the microscopic diatomaceæ to which I last evening referred, and of one shell and one coral, not a single species crossed the gap. On its further or remoter side, however, where the Secondary division closes, the intermingling of species again begins, and runs on till the commencement of this great Secondary division; and then, just where the Palæozoic division closes, we find another abrupt break, crossed, if crossed at all, — for there still exists some doubt on the subject, — by but two species of plant."

These breaks are considered to imply actual new creations on the surface of our planet; not only by Hugh Miller, but by the majority of geologists. And the terms Palæozoic, Mesozoic, and Cainozoic, are used to indicate these three successive systems of life. It is true that some accept this belief with caution: knowing how geologic research has been all along tending to fill up what were once thought wide breaks. Sir Charles Lyell points out that "the hiatus which exists in Great Britain between the fossils of the Lias and these of the Magnesian Limestone, is supplied in Germany by the rich fauna and flora of the Muschelkalk, Keuper, and Bunter Sandstein, which we know to be of a date precisely intermediate." Again he remarks that "until lately the fossils of the coal-measures were separated from those of the antecedent Silurian group by a very abrupt and decided line of demarcation; but recent discoveries have brought to light in Devonshire, Belgium, the Eifel, add Westphalia, the remains of a fauna of an intervening period." And once more, "we have also in like manner had some success of late years in diminishing the hiatus which still separates the Cretaceous and Eocene periods in Europe." To which let us add that since Hugh Miller penned the passage above quoted, the second of the great gaps he refers to has been very considerably narrowed by the discovery of strata containing Palæozoic genera and Mesozoic genera intermingled. Nevertheless, the occurrence of two great revolutions in the Earth's Flora and Fauna appears still to be held by many; and geologic nomenclature habitually assumes it.

Before seeking a solution of these phenomena, let us glance at the several minor causes that produce breaks in the geological succession of organic forms: taking first, the more general ones which modify climate, and, therefore, the distribution of life. Among these may be noted one which has not, we believe, been named by writers on the subject. We mean that resulting from a certain slow astronomic rhythm, by which the northern and southern hemispheres are alternately subject to greater extremes of temperature. In consequence of the slight ellipticity of its orbit, the Earth's distance from the sun varies to the extent of some 3,000,000 of miles. At present, the aphelion occurs at the time of our northern summer; and the perihelion during the summer of the southern hemisphere. In consequence, however, of that slow movement of the Earth's axis which produces the precession of the equinoxes, this state of things will in time be reversed: the Earth will be nearest to the sun during the summer of the northern hemisphere, and furthest from it during the southern summer or northern winter. The period required to complete the slow movement producing these changes, is nearly 26,000 years; and were there no modifying process, the two hemispheres would alternately experience this coincidence of summer with the least distance from the sun, during a period of 13,000 years. But there is also a still slower change in the direction of the axis major of the Earth's orbit; from which it results that the alternation we have described is completed in about 21,000 years. That is to say, if at a given time the Earth is nearest to the sun at our midsummer, and furthest from the sun at our mid-winter: then, in 10,500 years afterwards, it will be furthest from the sun at our mid-summer, and nearest at our mid-winter.

Now the difference between the distances from the sun at the two extremes of this alternation, amounts to one-thirtieth; and hence, the difference between the quantities of heat received from the sun on a summer's day under these opposite conditions amounts to one-fifteenth. Estimating this, not with reference to the zero of our thermometers, but with reference to the temperature of the celestial spaces, Sir John Herschel calculates "23° Fahrenheit as the least variation of temperature under such circumstances which can reasonably be attributed to the actual variation of the sun's distance." Thus, then, each hemisphere has at a certain epoch, a short summer of extreme heat, followed by a long and very cold winter. Through the slow change in the direction of the Earth's axis, these extremes are gradually mitigated. And at the end of 10,500 years, there is reached the opposite state — a long and moderate summer, with a short and mild winter. At present, in consequence of the predominance of sea in the southern hemisphere, the extremes to which its astronomical conditions subject it, are much ameliorated; while the great proportion of land in the northern hemisphere, tends to exaggerate such contrast as now exists in it between winter and summer: whence it results that the climates of the two hemispheres are not widely unlike. But 10,000 years hence, the northern hemisphere will undergo annual variations of temperature far more marked than now.

In the last edition of his *Outlines of Astronomy,* Sir John Herschel recognizes this as an element in geological processes: regarding it as possibly a part-cause of those climatic changes indicated by the records of the Earth's past. That it has had much to do with the larger changes of climate of which we have evidence, seems unlikely, since there is reason to think that these have been far slower and more lasting; but that it must have entailed a rhythmical exaggeration and mitigation of the climates otherwise produced, seems beyond question. And it seems also beyond question that there must have been a consequent rhythmical change in the distribution of organisms — a rhythmical change to which we here wish to draw attention, as one cause of minor breaks in the succession of fossil remains. Each species of plant and animal, has certain limits of heat and cold within which only it can exist; and these limits in a great degree determine its geographical position. It will not spread north of a certain latitude, because it cannot bear a more northern winter, nor south of a certain latitude, because the summer heat is too great; or else it is indirectly restrained from spreading further by the effect of temperature on the humidity of the air, or on the distribution of the organisms it lives upon.

But now, what will result from a slow alteration of climate, produced as above described? Supposing the period we set out from is that in which the contrast of seasons is least marked, it is manifest that during the progress towards the period of the most violent contrast, each species of plant and animal will gradually change its limits of distribution — will be driven back, here by the winter's increasing cold, and there by the summer's increasing heat — will retire into those localities that are still fit for it. Thus during 10,000 years, each species will ebb away from certain regions it was inhabiting; and during the succeeding 10,000 years will flow back into those regions. From the strata there forming, its remains will disappear; they will be absent from some of the supposed strata; and will be found in strata higher up. But in what shapes will they re-appear? Exposed during the 21,000 years of their slow recession and their slow return, to changing conditions of life, they are likely to have undergone modifications; and will probably re-appear with slight differences of constitution and perhaps of form — will be new varieties or perhaps new sub-species.

To this cause of minor breaks in the succession of organic forms — a cause on which we have dwelt because it has not been taken into account — we must add sundry others. Besides these periodically-recurring alterations of climate, there are the irregular ones produced by re-distributions of land and sea; and these, sometimes less, sometimes greater, in degree, than the rhythmical changes, must, like them, cause in each region the ebb and flow of species; and consequent breaks, small or large as the case may be, in the palæontological series. Other and more special geological changes must produce other and more local blanks in the succession of fossils. By some inland elevation the natural drainage of a continent is modified; and instead of the sediment it previously brought down to the sea, a great river begins to bring down sediment unfavourable to various plants and animals living in its delta: wherefore these disappear from the locality, perhaps to re-appear in a changed form after a long epoch. Upheavals or subsidences of shores or sea-bottoms, involving deviations of marine currents, must remove the habitats of many species to which such currents are salutary or injurious; and further, this re-distribution of currents must alter the places of sedimentary deposits, and so stop the burying of organic remains in some localities, and commence it in others. Had we space, many more such causes of blanks in our palæontological records might be added. But it is needless here to enumerate them. They are admirably explained and illustrated in Sir Charles Lyell's *Principles of Geology.*

Now, if these minor revolutions of the Earth's surface produce minor breaks in the series of fossilized remains; must not great revolutions produce great breaks? If a local upheaval or subsidence causes throughout its small area the absence of some links in the chain of fossil forms; does it not follow that an upheaval or subsidence extending over a large part of the Earth's surface, must cause the absence of a great number of such links throughout a very wide area?

When during a long epoch a continent, slowly subsiding, gives place to a far-spreading ocean some miles in depth, at the bottom of which no deposits from rivers or abraded shores can be thrown down; and when, after some enormous period, this ocean-bottom is gradually elevated and becomes the site of new strata; it is clear that the fossils contained in these new strata are likely to have but little in common with the fossils of the strata below them. Take, in illustration, the case of the North Atlantic. We have already named the fact that between this country and the United States, the ocean-bottom is being covered with a deposit of chalk — a deposit that has been forming, probably, ever since there occurred that great depression of the Earth's crust from which the Atlantic resulted in remote geologic times. This chalk consists of the minute shells of Foraminifera, sprinkled with remains of small Entomostraca, and probably a few Pteropod-shells: though the sounding lines have not yet brought up any of these last. Thus, in so far as all high forms of life are concerned, this new chalk-formation must be a blank. At rare intervals, perhaps, a polar bear drifted on an iceberg, may have its bones scattered over the bed; or a dead, decaying whale may similarly leave traces. But such remains must be so rare, that this new chalk-formation, if visible, might be examined for a century before any of them were disclosed. If now, some millions of years hence, the Atlantic-bed should be raised, and estuary or shore deposits laid upon it, these deposits would contain remains of a Flora and Fauna so distinct from everything below them, as to appear like a new creation.

Thus, along with continuity of life on the Earth's surface, there not only *may* be, but there *must* be, great gaps, in the series of fossils; and hence these gaps are no evidence against the doctrine of Evolution.

One other current assumption remains to be criticized; and it is the one on which, more than on any other, depends the view taken respecting the question of development.

From the beginning of the controversy, the arguments for and against have turned upon the evidence of progression in organic forms, found in the ascending series of our sedimentary formations. On the one hand, those who contend that higher organisms have been evolved out of lower, joined, with those who contend that successively higher organisms have been created at successively later periods, appeal for proof to the facts of Palæontology; which, they say, countenance their views. On the other hand, the Uniformitarians, who not only reject the hypothesis of development, but deny that the modern forms of life are higher than the ancient ones, reply that the Palæontological evidence is at present very incomplete; that though we have not yet found remains of highly-organized creatures in strata of the greatest antiquity, we must not assume that no such creatures existed when those strata were deposited; and that, probably, geological research will eventually disclose them.

It must be admitted that thus far, the evidence has gone in favour of the latter party. Geological discovery has year after year shown the small value of negative facts. The conviction that there are no traces of higher organisms in earlier strata, has resulted not from the absence of such remains, but from incomplete examination. At p. 460 of his *Manual of Elementary Geology,* Sir Charles Lyell gives a list in illustration of this. It appears that in 1709, fishes were not known lower than the Permian system. In 1793 they were found in the subjacent Carboniferous system; in 1828 in the Devonian; in 1840 in the Upper Silurian. Of reptiles, we read that in 1710 the lowest known were in the Permian; in 1844 they were detected in the Carboniferous; and in 1852 in the Upper Devonian. While of the Mammalia the list shows that in 1798 none had been discovered below the middle Eocene; but that in 1818 they were discovered in the Lower Oolite; and in 1847 in the Upper Trias.

The fact is, however, that both parties set out with an inadmissible postulate. Of the Uniformitarians, not only such writers as Hugh Miller, but also such as Sir Charles Lyell,[20] reason as though we had found the earliest, or something like the earliest, strata. Their antagonists, whether defenders of the Development Hypothesis or simply Progressionists, almost uniformly do the like. Sir R. Murchison, who is a Progressionist, calls the lowest fossiliferous strata, "Protozoic." Prof. Ansted uses the same term. Whether avowedly or not, all the disputants stand on this assumption as their common ground.

Yet is this assumption indefensible, as some who make it very well know. Facts may be cited against it which show that it is a more than questionable one — that it is a highly improbable one; while the evidence assigned in its favour will not bear criticism.

Because in Bohemia, Great Britain, and portions of North

---

[20] Sir Charles Lyell is no longer to be classed among Uniformitarians. With rare and admirable candour he has, since this was written, yielded to the arguments of Mr. Darwin.

America, the lowest unmetamorphosed strata yet discovered, contain but slight traces of life, Sir R. Murchison conceives that they were formed while yet few, if any, plants or animals had been created; and, therefore, classes them as "Azoic." His own pages, however, show the illegitimacy of the conclusion that there existed at that period no considerable amount of life. Such traces of life as have been found in the Longmynd rocks, for many years considered unfossiliferous, have been found in some of the lowest beds; and the twenty thousand feet of superposed beds, still yield no organic remains. If now these superposed strata throughout a depth of four miles, are without fossils, though the strata over which they he prove that life had commenced; what becomes of Sir R. Murchison's inference? At page 189 of *Siluria,* a still more conclusive fact will be found. The "Glengariff grits," and other accompanying strata there described as 13,500 feet thick, contain no signs of contemporaneous life. Yet Sir R. Murchison refers them to the Devonian period — a period that had a large and varied marine Fauna. How then, from the absence of fossils in the Longmynd beds and their equivalents, can we conclude that the Earth was "azoic" when they were formed?

"But," it may be asked, "if living creatures then existed, why do we not find fossiliferous strata of that age, or an earlier age?" One reply is, that the non-existence of such strata is but a negative fact — we have not found them. And considering how little we know even of the two-fifths of the Earth's surface now above the sea, and how absolutely ignorant we are of the three-fifths below the sea, it is rash to say that no such strata exist. But the chief reply is, that these records of the Earth's earlier history have been in great part destroyed, by agencies that are eve tending to destroy such records.

It is an established geological doctrine, that sedimentary strata are liable to be changed, more or less completely, by igneous action. The rocks originally classed as "transition," because they were intermediate in character between the igneous rocks found below them, and the sedimentary strata found above them, are now known to be nothing else than sedimentary strata altered in texture and appearance by the intense heat of adjacent molten matter; and hence are renamed "metamorphic rocks." Modern researches have shown, too, that these metamorphic rocks are not, as was once supposed, all of the same age. Besides primary and secondary strata that have been transformed by igneous action, there are similarly-changed deposits of tertiary origin; and that, even for a quarter of a mile from the point of contact with neighbouring granite. By this process fossils are of course destroyed. "In some cases," says Sir Charles Lyell, "dark limestones, replete with shells and corals, have been turned into white statuary marble, and hard clays, containing vegetable or other remains, into slates called mica-schist or hornblende-schist; every vestige of the organic bodies having been obliterated."

Again, it is fast becoming an acknowledged truth, that igneous rock, of whatever kind, is the product of sedimentary strata that have been completely melted. Granite and gneiss, which are of like chemical composition, have been shown, in various cases, to pass one into the other: as at Valorsine, near Mont Blanc, where the two, in contact, are observed to "both undergo a modification of mineral character. The granite still remaining unstratified, becomes charged with green particles; and the talcose gneiss assumes a granitiform structure without losing its stratification." In the Aberdeen-granite, lumps of unmelted gneiss are frequently found; and we can ourselves bear witness that on the banks of Loch Sunart, there is ample proof that the granite of that region, when it was molten, contained incompletely-fused clots of sedimentary strata. Nor is this all. Fifty years ago, it was thought that all granitic rocks were primitive, or existed before any sedimentary strata; but it is now "no easy task to point out a single mass of granite demonstrably more ancient than all the known fossiliferous deposits."

In brief, accumulated evidence clearly shows, that by contact with, or proximity to, the molten matter of the Earth's nucleus, all beds of sediment are liable to be actually melted, or partially fused, or so heated as to agglutinate their particles; and that according to the temperature they have been raised to, and the circumstances under which they cool, they assume the forms of granite, porphyry, trap, gneiss, or rock otherwise altered. Further, it is manifest that though strata of various ages have been thus changed, yet that the most ancient strata have been as changed to the greatest extent: both because they have habitually lain nearer to the centre of igneous agency; and because they have been for a longer period liable to the effects of this agency.

Whence it follows, that sedimentary strata passing a certain antiquity, are unlikely to be found in an unmetamorphosed state; and that strata much earlier than those are certain to have been melted up. Thus if, throughout a past of indefinite duration, there had been at work those aqueous and igneous agencies which we see still at work, the state of the Earth's crust might be just what we find it. We have no evidence which puts a limit to the period throughout which this formation and destruction of strata has been going on. For aught the facts prove, it may have been going on for ten times the period measured by our whole series of sedimentary deposits.

Besides having, in the present appearances of the Earth's crust, no data for fixing a commencement to these processes — besides finding that the evidence permits us to assume such commencement to have been inconceivably remote, as compared even with the vast eras of geology; we are not without positive grounds for inferring the inconceivable remoteness of such commencement. Modern geology has established truths which are irreconcilable with the belief that the formation and destruction of strata began when the Cambrian rocks were formed; or at anything like so recent a time. One fact from *Siluria* will suffice. Sir R. Murchison estimates the vertical thickness of Silurian strata in Wales, at from 26,000 to 27,000 feet, or about five miles; and if to this we add the vertical depth of the Cambrian strata, on which the Silurians lie conformably, there results, on the lowest computation, a total depth of seven miles.

Now it is held by geologists, that this vast accumulation of strata must have been deposited in an area of gradual subsidence. These strata could not have been thus laid on each other in regular order, unless the Earth's crust had been at that place sinking, either continuously or by very small steps. Such an immense subsidence, however, must have been impossible without a crust of great thickness. The Earth's molten nucleus tends ever, with enormous force, to assume the form of a regular oblate spheroid. Any depression of its crust below the surface of equilibrium, and any elevation of its crust above that surface, have to withstand immense resistance. It follows inevitably that, with a thin crust, nothing but small elevations and subsidences would be possible; and that, conversely, a subsidence of seven miles implies a crust of comparatively great strength, or, in other words, of great thickness. Indeed, if we compare this inferred subsidence in the Silurian period, with such elevations and depressions as our existing continents and oceans display, we see no evidence that the Earth's crust was appreciably thinner then than now. What are the implications? If, as geologists generally admit, the Earth's crust has resulted from that slow cooling which is even still going on — if we see no sign that at the time when the earliest Cambrian strata were formed, this crust was appreciably thinner than now; we are forced to conclude that the era during which it acquired that great thickness possessed in the Cambrian period, was enormous as compared with the interval between the Cambrian period and our own. But during the incalculable series of epochs thus inferred, there existed an ocean, tides, winds, waves, rain, rivers. The agencies by which the denudation of continents and filling up of seas have all along been carried on, were as active then as now. Endless successions of strata must have been formed. And when we ask — Where are they? Nature's obvious reply is — They have been destroyed by that igneous action to which so great a part of our oldest-known strata owe their fusion or metamorphosis.

Only the last chapter of the Earth's history has come down to us. The many previous chapters, stretching back to a time immeasurably remote, have been burnt; and with them all the records of life we may presume they contained. The greater part of the evidence which might have served to settle the Development-controversy, is for ever lost; and on neither side can the arguments derived from Geology be conclusive.

"But how happen there to be such evidences of progression as exist?" it may be asked. "How happens it that, in ascending from the most ancient strata to the most recent strata, we do find a succession of organic forms, which, however irregularly, carries us from lower to higher?" This question seems difficult to answer. Nevertheless, there is reason for thinking that nothing can be safely inferred from the apparent progression here cited. And the illustration which shows as much, will, we believe, also show how little trust is to be placed in certain geological generalizations that appear to be well established. With this somewhat elaborate illustration, to which we now pass, our criticisms may fitly conclude.

Let us suppose that in a region now covered by wide ocean,

there begins one of those great and gradual upheavals by which new continents are formed. To be precise, let us say that in the South Pacific, midway between New Zealand and Patagonia, the sea-bottom has been little by little thrust up towards the surface, and is about to emerge. What will be the successive phenomena, geological and biological, which are likely to occur before this emerging sea-bottom has become another Europe or Asia?

In the first place, such portions of the incipient land as are raised to the level of the waves, will be rapidly denuded by them: their soft substance will be torn up by the breakers, carried away by the local currents, and deposited in neighbouring deeper water. Successive small upheavals will bring new and larger areas within reach of the waves; fresh portions will each time be removed from the surfaces previously denuded; and further, some of the newly-formed strata, being elevated nearly to the level of the water, will be washed away and re-deposited. In course of time, the harder formations of the upraised sea-bottom will be uncovered. These being less easily destroyed, will remain permanently above the surface; and at their margins will arise the usual breaking down of rocks into beach-sand and pebbles. While in the slow process of this elevation, going on at the rate of perhaps two or three feet in a century, most of the sedimentary deposits produced will be again and again destroyed and re-formed; there will, in those adjacent areas of subsidence which accompany areas of elevation, be more or less continuous successions of sedimentary deposits.

And now what will be the character of these new strata? They will necessarily contain scarcely any traces of life. The deposits that had previously been slowly formed at the bottom of this wide ocean, would be sprinkled with fossils of but few species. The oceanic Fauna is not a rich one; its hydrozoa do not admit of preservation; and the hard parts of its few kinds of molluscs and crustaceans and insects are mostly fragile. Hence, when the ocean-bed was here and there raised to the surface — when its strata of sediment with their contained organic fragments were torn up and long washed about by the breakers before being re-deposited — when the re-deposits were again and again subject to this violent abrading action by subsequent small elevations, as they would mostly be; what few fragile organic remains they contained, would be in nearly all cases destroyed.

Thus such of the first-formed strata as survived the repeated changes of level, would be practically "azoic;" like the Cambrian of our geologists. When by the washing away of the soft deposits, the hard sub-strata had been exposed in the shape of rocky islets, and a footing had thus been furnished, the pioneers of a new life might be expected to make their appearance. What would they be? Not any of the surrounding oceanic species, for these are not fitted for a littoral life; but species flourishing on some of the far-distant shores of the Pacific. Of such the first to establish themselves would be sea-weeds and zoophytes; both because their swarming spores and gemmules would be the most readily conveyed with safety, and because when conveyed they would find fit food. It is true that Cirrhipeds and Lamellibranchs, subsisting on the minute creatures which everywhere people the sea, would also find fit food.

But passing over the fact that the germs of such higher forms are neither so abundant nor so well fitted to bear long voyages, there is the more important fact that the individuals arising from these germs can reproduce only sexually, and that this vastly increases the obstacles to the establishment of their races. The chances of early colonization are immensely in favour of species which, multiplying by agamogenesis, can people a whole shore from a single germ; and immensely against species which, multiplying only by gamogenesis, must be introduced in considerable numbers that some may survive, meet, and propagate. Thus we infer that the earliest traces of life left in the sedimentary deposits near these new shores, will be traces of life as humble as that indicated in the most ancient rocks of Great Britain and Ireland. Imagine now that the processes we have briefly indicated, continue — that the emerging lands become wider in extent, and fringed by higher and more varied shores; and that there still go on those ocean-currents which, at long intervals, convey from far distant shores immigrant forms of life. What will result? Lapse of time will of course favour the introduction of such new forms: admitting, as it must, of those combinations of fit conditions, which, under the law of probabilities, can occur only at very distant intervals. Moreover, the increasing area of the islands, individually and as a group, implies increasing length of coast; from which there follows a longer line of contact with the streams and waves that bring drifting masses; and, therefore, a greater chance that germs of fresh life will be stranded.

And once more, the comparatively-varied shores, presenting physical conditions that change from mile to mile, will furnish suitable habitats for more numerous species. So that as the elevation proceeds, three causes conspire to introduce additional marine plants and animals. To what classes will the increasing Fauna be for a long period confined? Of course, to classes of which individuals, or their germs, are most liable to be carried far away from their native shores by floating sea-weed or drift-wood; to classes which are also least likely to perish in transit, or from change of climate; and to those which can best subsist around coasts comparatively bare of life. Evidently, then, corals, annelids, inferior molluscs, and crustaceans of low grade, will chiefly constitute the early Fauna. The large predatory members of these classes, will be later in establishing themselves; both because the new shores must first become well peopled by the creatures they prey on, and because, being more complex, they or their ova must be less likely to survive the journey, and the change of conditions.

We may infer, then, that the strata deposited next after the almost "azoic" strata, would contain the remains of invertebrata, allied to those found near the shores of Australia and South America. Of such invertebrate remains, the lower beds would furnish comparatively few genera, and those of relatively low types; while in the upper beds the number of genera would be greater, and the types higher: just as among the fossils of our Silurian system. As this great geologic change slowly progressed through its long history of earthquakes, volcanic disturbances, minor upheavals and subsidences — as the extent of the archipelago became greater and its smaller islands coalesced into larger ones, while its coast line grew still longer and more varied, and the neighbouring sea more thickly inhabited by inferior forms of life; the lowest division of the vertebrata would begin to be represented. In order of time, fish would naturally come after the lower invertebrata: both as being less likely to have their ova transported across the waste of waters, and as requiring for their subsistence a pre-existing Fauna of some development. They might be expected to make their appearance along with the predaceous crustaceans; as they do in the uppermost Silurian rocks.

And here, too, let us remark, that as, during this long epoch we have been describing, the sea would have made great inroads on some of the newly raised lands that had remained stationary; and would probably in some places have reached masses of igneous or metamorphic rocks there might, in course of time, arise by the decomposition and denudation of such rocks, local deposits coloured with oxide of iron, like our Old Red Sandstone. And in these deposits might be buried the remains of the fish then peopling the neighbouring sea.

Meanwhile, how would the surfaces of the upheaved masses be occupied? At first their deserts of naked rocks and pebbles would bear only the humblest forms of vegetal life, such as we find in grey and orange patches on our own rugged mountain sides; for these alone could flourish on such surfaces, and their spores would be the most readily transported. When, by the decay of such protophytes, and that decomposition of rock effected by them, there had resulted a fit habitat for mosses; these, of which the germs might be conveyed in drifted trees, would begin to spread. A soil having been eventually thus produced, it would become possible for plants of higher organization to find root-hold; and as in the way we have described the archipelago and its constituent islands grew larger, and had more multiplied relations with winds and waters, such higher plants might be expected ultimately to have their seeds transferred from the nearest lands. After something like a Flora had thus colonized the surface, it would become possible for insects to exist; and of air-breathing creatures, insects would manifestly be among the first to find their way from elsewhere.

As, however, terrestrial organisms, both vegetal and animal, are much less likely than marine organisms to survive the accidents of transport from distant shores; it is clear that long after the sea surrounding these new lands had acquired a varied Flora and Fauna, the lands themselves would still be comparatively bare; and thus that the early strata, like our Silurians, would afford no traces of terrestrial life. By the time that large areas had been raised above the ocean, we may fairly suppose a luxuriant vegetation to have been acquired. Under what circumstances are we likely to find this vegetation fossilized? Large surfaces of land imply large rivers with their accompanying deltas; and are liable to have lakes and swamps. These, as we know from extant cases, are favourable to rank vegetation; and afford the conditions needful for preserving it in the shape of coal-beds. Observe, then, that while in the early history of such a continent a carboniferous period could not occur, the occurrence of a carboniferous period would become probable after long-continued upheavals had uncovered large areas. As in our own sedimentary series, coal-beds would make their appearance only after there had been enormous accumulations of earlier strata charged with marine fossils.

Let us ask next, in what order the higher forms of animal life would make their appearance. We have seen how, in the succession of marine forms, there would be something like a progress from the lower to the higher: bringing us in the end to predaceous molluscs, crustaceans, and fish. What are likely to succeed fish? After marine creatures, those which would have the greatest chance of surviving the voyage would be amphibious reptiles: both because they are more tenacious of life than higher animals, and because they would be less completely out of their element. Such reptiles as can live in both fresh and salt water, like alligators; and such as are drifted out of the mouths of great rivers on floating trees, as Humboldt says the Orinoco alligators are; might be early colonists.

It is manifest, too, that reptiles of other kinds would be among the first vertebrata to people the new continent. If we consider what will occur on one of those natural rafts of trees, soil, and matted vegetable matter, sometimes swept out to sea by such currents as the Mississippi, with a miscellaneous living cargo; we shall see that while the active, hot-blooded, highly-organized creatures will soon die of starvation and exposure, the inert, cold-blooded ones, which can go long without food, will live perhaps for weeks; and so, out of the chances from time to time occurring during long periods, reptiles will be the first to get safely lauded on foreign shores: as indeed they are even now known sometimes to be. The transport of mammalia being comparatively precarious, must, in the order of probability, be longer postponed; and would, indeed, be unlikely to occur until by the enlargement of the new continent, the distances of its shores from adjacent lands had been greatly diminished, or the formation of intervening islands had increased the chances of survival.

Assuming, however, that the facilities of immigration had

become adequate; which would be the first mammals to arrive and live? Not large herbivores; for they would be soon drowned if by any accident carried out to sea. Not the carnivora; for these would lack appropriate food, even if they outlived the voyage. Small quadrupeds frequenting trees, and feeding on insects, would be those most likely both to be drifted away from their native lands and to find fit food in a new one. Insectivorous mammals, like in size to those found in the Trias and the Stonesfield slate, might naturally be looked for as the pioneers of the higher vertebrata. And if we suppose the facilities of communication to be again increased, either by a further shallowing of the intervening sea and a consequent multiplication of islands, or by an actual junction of the new continent with an old one, through continued upheavals; we should finally have an influx of the larger and more perfect mammals.

Now rude as is this sketch of a process that would be extremely elaborate and involved, and open as some of its propositions are to criticisms which there is no space here to meet; no one will deny that it represents something like the biologic history of the supposed new continent. Details apart, it is manifest that simple organisms, able to flourish under simple conditions of life, would be the first successful immigrants; and that more complex organisms, needing for their existence the fulfilment of more complex conditions, would afterwards establish themselves in something like an ascending succession. At the one extreme we see every facility. The new individuals can be conveyed in the shape of minute germs; these are infinite in their numbers; they are diffused in the sea; they are perpetually being carried in all directions to great distances by ocean-currents; they can survive such long journeys unharmed; they can find nutriment wherever they arrive; and the resulting organisms can multiply asexually with great rapidity.

At the other extreme, we see every difficulty. The new individuals must be conveyed in their adult forms; their numbers are, in comparison, utterly insignificant; they live on land, and are very unlikely to be carried out to sea; when so carried, the chances are immense against their escape from drowning, starvation, or death by cold; if they survive the transit, they must have a pre-existing Flora or Fauna to supply their special food; they require, also, the fulfilment of various other physical conditions; and unless at least two individuals of different sexes are safely landed, the race cannot be established. Manifestly, then, the immigration of each successively higher order of organisms, having, from one or other additional condition to be fulfilled, an enormously-increased probability against it, would naturally be separated from the immigration of a lower order by some period like a geologic epoch.

And thus the successive sedimentary deposits formed while this new continent was undergoing gradual elevation, would seem to furnish clear evidence of a general progress in the forms of life. That lands thus raised up in the midst of a wide ocean, would first give origin to unfossiliferous strata; next, to strata containing only the lowest marine forms; next, to strata containing higher marine forms, ascending finally to fish; and that the strata above these would contain reptiles, then small mammals, then great mammals; seems to us to be demonstrable from the known laws of organic life.

And if the succession of fossils presented by the strata of this supposed new continent, would thus simulate the succession presented by our own sedimentary series; must we not say that our own sedimentary series very possibly records nothing more than the phenomena accompanying one of these great upheavals? We think this must be considered not only possible, but highly probable: harmonizing as it does with the unavoidable conclusion before pointed out, that geological changes must have been going on for a period immeasurably greater than that of which we have records. And if the probability of this conclusion be admitted, it must be admitted that the facts of Palæontology can never suffice either to prove or disprove the Development Hypothesis; but that the most they can do is, to show whether the last few pages of the Earth's biologic history are or are not in harmony with this hypothesis — whether the existing Flora and Fauna can or can not be affiliated upon the Flora and Fauna of the most recent geologic times.

IX
THE DEVELOPMENT HYPOTHESIS

In a debate upon the development hypothesis, lately narrated to me by a friend, one of the disputants was described as arguing, that as, in all our experience, we know no such phenomenon as transmutation of species, it is unphilosophical to assume that transmutation of species ever takes place. Had I been present, I think that, passing over his assertion, which is open to criticism, I should have replied that, as in all our experience we have never known a species *created,* it was, by his own showing, unphilosophical to assume that any species ever had been created.

Those who cavalierly reject the Theory of Evolution, as not adequately supported by facts, seem quite to forget that their own theory is supported by no facts at all. Like the majority of men who are born to a given belief, they demand the most rigorous proof of any adverse belief, but assume that their own needs none. Here we find, scattered over the globe, vegetable and animal organisms numbering, of the one kind (according to Humboldt), some 320,000 species, and of the other, some 2,000,000 species (see Carpenter); and if to these we add the numbers of animal and vegetable species that have become extinct, we may safely estimate the number of species that have existed, and are existing, on the Earth, at not less than *ten millions.* Well, which is the most rational theory about these ten millions of species? Is it most likely that there have been ten millions of special creations? or is it most likely that by continual modifications, due to change of circumstances, ten millions of varieties have been produced, as varieties are being produced still?

Doubtless many will reply that they can more easily conceive ten millions of special creations to have taken place, than they can conceive that ten millions of varieties have arisen by successive modifications. All such, however, will find, on inquiry, that they are under an illusion. This is one of the many cases in which men do not really believe, but rather *believe they believe.* It is not that they can truly conceive ten millions of special creations to have taken place, but that they *think they can do so.* Careful introspection will show them that they have never yet realized. to themselves the creation of even *one* species. If they have formed a definite conception of the process, let them tell us how a new species is constructed, and how it makes its appearance. Is it thrown down from the clouds? or must we hold to the notion that it struggles up out of the ground? Do its limbs and viscera rush together from all the points of the compass? or must we receive the old Hebrew idea, that God takes clay and moulds a new creature? If they say that a new creature is produced in none of these modes, which are too absurd to be believed; then they are required to describe the mode in which a new creature *may* be produced — a mode which does *not* seem absurd: and such a mode they will find that they neither have conceived nor can conceive.

Should the believers in special creations consider it unfair thus to call upon them to describe how special creations take place, I reply, that this is far less than they demand from the supporters of the Development Hypothesis. They are merely asked to point out a *conceivable* mode. On the other hand, they ask, not simply for a *conceivable* mode, but for the *actual* mode. They do not say — Show us how this *may* take place; but they say — Show us how this *does* take place. So far from its being unreasonable to put the above question, it would be reasonable to ask not only for a *possible* mode of special creation, but for an *ascertained* mode; seeing that this is no greater a demand than they make upon their opponents.

And here we may perceive how much more defensible the new doctrine is than the old one. Even could the supporters of the Development Hypothesis merely show that the origination of species by the process of modification is conceivable, they would be in a better position than their opponents. But they can do much more than this. They can show that the process of modification has effected, and is effecting, decided changes in all organisms subject to modifying influences. Though, from the impossibility of getting at a sufficiency of facts, they are unable to trace the many phases through which any existing species has passed in arriving at its present form, or to identify the influences which caused the successive modifications; yet, they can show that any existing species — animal or vegetable — when placed under conditions different from its previous ones, *immediately begins to undergo certain changes of structure fitting it for the new conditions.* They can show that in successive generations these changes continue, until ultimately the new conditions become the natural ones. They can show that in cultivated plants, in domesticated animals, and in the several races of men, such alterations have taken place. They can show that the degrees of difference so produced are often, as in dogs, greater than those on which distinctions of species are in other cases founded. They can show that it is a matter of dispute whether some of these modified forms *are* varieties or separate species. They can show, too, that the changes daily taking place in ourselves — the facility that attends long practice, and the loss of aptitude that begins when practice ceases — the strengthening of passions habitually gratified, and the weakening of those habitually curbed — the development of every faculty, bodily, moral, or intellectual, according to the use made of it — are all explicable on this same principle. And thus they can show that throughout all organic nature there *is* at work a modifying influence of the kind they assign as the cause of these specific differences: an influence which, though slow in its action, does, in time, if the circumstances demand it, produce marked changes — an influence which, to all appearance, would produce in the millions of years, and under the great varieties of condition which geological records imply, any amount of change.

Which, then, is the most rational hypothesis? — that of special creations which has neither a fact to support it nor is even definitely conceivable; or that of modification, which is not only definitely conceivable, but is countenanced by the habitudes of every existing organism?

That by any series of changes a protozoon should ever become a mammal, seems to those who are not familiar with zoology, and who have not seen how clear becomes the relationship between the simplest and the most complex forms when intermediate forms are examined, a very grotesque notion. Habitually looking at things rather in their statical than in their dynamical aspect, they never realize the fact that, by small increments of modification, any amount of modification may in time be generated. That surprise which they feel on finding one whom they last saw as a boy, grown into a man, becomes incredulity when the degree of change is greater. Nevertheless, abundant instances are at hand of the mode in which we may pass to the most diverse forms, by insensible gradations. Arguing the matter some time since with a learned professor, I illustrated my position thus: — You admit that there is no apparent relationship between a circle and an hyperbola. The one is a finite curve; the other is an infinite one. All parts of the one are alike; of the other no two parts are alike. The one incloses a space; the other will not inclose a space though produced for ever. Yet opposite as are these curves in all their properties, they may be connected together by a series of intermediate curves, no one of which differs from the adjacent ones in any appreciable degree. Thus, if a cone be cut by a plane at right angles to its axis we get a circle. If, instead of being perfectly at right angles, the plane subtends with the axis an angle of 89° 59', we have an ellipse, which no human eye, even when aided by an accurate pair of compasses, can distinguish from a circle. Decreasing the angle minute by minute, the ellipse becomes first perceptibly eccentric, then manifestly so, and by and by acquires so immensely elongated a form, as to bear no recognisable resemblance to a circle. By continuing this process, the ellipse passes insensibly into a parabola; and ultimately, by still further diminishing the angle, into an hyperbola. Now here we have four different species of curve — circle, ellipse, parabola, and hyperbola — each having its peculiar properties and its separate equation, and the first and last of which are quite opposite in nature, connected together as members of one series, all producible by a single process of insensible modification.

But the blindness of those who think it absurd to suppose that complex organic forms may have arisen by successive modifications out of simple ones, becomes astonishing when we remember that complex organic forms are daily being thus produced. A tree differs from a seed immeasurably in every respect — in bulk, in structure, in colour, in form, in specific gravity, in chemical composition: differs so greatly that no visible resemblance of any kind can be pointed out between them. Yet is the one changed in the course of a few years into the other: changed so gradually, that at no moment can it be said — Now the seed ceases to be, and the tree exists. What can be more widely contrasted than a newly-born child and the small, semi-transparent, gelatinous spherule constituting the human ovum? The infant is so complex in structure that a cyclopædia is needed to describe its constituent parts. The germinal vesicle is so simple that it may be defined in a line. Nevertheless, a few months suffice to develop the one out of the other; and that, too, by a series of modifications so small, that were the embryo examined at successive minutes, even a microscope would with difficulty disclose any sensible changes. That the uneducated and the ill-educated should think the hypothesis that all races of beings, man inclusive, may in process of time have been evolved from the simplest monad, a ludicrous one, is not to be wondered at. But for the physiologist, who knows that every individual being is so evolved — who knows further, that in their earliest condition the germs of all plants and animals whatever are so similar, "that there is no appreciable distinction amongst them which would enable it to be determined whether a particular molecule is the germ of a conferva or of an oak, of a zoophyte or of a man;"[21] — for him to make a difficulty of the matter is inexcusable. Surely if a single cell may, when subjected to certain influences, become a man in the space of twenty years; there is nothing absurd in the hypothesis that under certain other influences, a cell may in the course of millions of years give origin to the human race. The two processes are generically the same; and differ only in length and complexity.

We have, indeed, in the part taken by many scientific men in this controversy of "Law *versus* Miracle," a good illustration of the tenacious vitality of superstitions. Ask one of our leading

---

[21] Carpenter.

geologists or physiologists whether he believes in the Mosaic account of the creation, and he will take the question as next to an insult. Either he rejects the narrative entirely, or understands it in some vague non-natural sense. Yet one part of it he unconsciously adopts; and that, too, literally. For whence has he got this notion of "special creations," which he thinks so reasonable, and fights for so vigorously? Evidently he can trace it back to no other source than this myth which he repudiates. He has not a single fact in nature to quote in proof of it; nor is he prepared with any chain of abstract reasoning by which it may be established. Catechise him, and he will be forced to confess that the notion was put into his mind in childhood as part of a story which he now thinks absurd. And why, after rejecting all the rest of this story, he should strenuously defend this last remnant of it as though he had received it on valid authority, he would be puzzled to say.

X
THE SOCIAL ORGANISM

Sir James Macintosh got great credit for the saying, that "constitutions are not made, but grow." In our day, the most significant thing about this saying is, that it was ever thought so significant. As from the surprise displayed by a man at some familiar fact, you may judge of his general culture; so from the admiration which an age accords to a new thought, its average degree of enlightenment may be inferred. That this apophthegm of Macintosh should have been quoted and re-quoted as it has, shows how profound has been the ignorance of social science. A small ray of truth has seemed brilliant, as a distant rushlight looks like a star in the surrounding darkness.

Such a conception could not, indeed, fail to be startling when let fall in the midst of a system of thought to which it was utterly alien. Universally in Macintosh's day, things were explained on the hypothesis of manufacture, rather than that of growth: as indeed they are, by the majority, in our own day. It was held that the planets were severally projected round the sun from the Creator's hand; with exactly the velocity required to balance the sun's attraction. The formation of the Earth, the separation of sea from land, the production of animals, were mechanical works from which God rested as a labourer rests. Man was supposed to be moulded after a manner somewhat akin to that in which a modeller makes a clay-figure. And of course, in harmony with such ideas, societies were tacitly assumed to be arranged thus or thus by direct interposition of Providence; or by the regulations of law-makers; or by both.

Yet that societies are not artificially put together, is a truth so manifest, that it seems wonderful men should have ever overlooked it. Perhaps nothing more clearly shows the small value of historical studies, as they have been commonly pursued. You need but to look at the changes going on around, or observe social organization in its leading peculiarities, to see that these are neither supernatural, nor are determined by the wills of individual men, as by implication historians commonly teach; but are consequent on general natural causes. The one case of the division of labour suffices to show this. It has not been by command of any ruler that some men have become manufacturers, while others have remained cultivators of the soil. In Lancashire, millions have devoted themselves to the making of cotton-fabrics; in Yorkshire, another million lives by producing woollens; and the pottery of Staffordshire, the cutlery of Sheffield, the hardware of Birmingham, severally occupy their hundreds of thousands. These are large facts in the structure of English society; but we can ascribe them neither to miracle, nor to legislation. It is not by "the hero as king," any more than by "collective wisdom," that men have been segregated into producers, wholesale distributors, and retail distributors.

The whole of our industrial organization, from its main outlines down to its minutest details, has become what it is, not simply without legislative guidance, but, to a considerable extent, in spite of legislative hindrances. It has arisen under the pressure of human wants and activities. While each citizen has been pursuing his individual welfare, and none taking thought about division of labour, or, indeed, conscious of the need for it, division of labour has yet been ever becoming more complete. It has been doing this slowly and silently: scarcely any having observed it until quite modern times. By steps so small, that year after year the industrial arrangements have seemed to men just what they were before — by changes as insensible as those through which a seed passes into a tree; society has become the complex body of mutually-dependent workers which we now see. And this economic organization, mark, is the all-essential organization. Through the combination thus spontaneously evolved, every citizen is supplied with daily necessaries; while he yields some product or aid to others. That we are severally alive to-day, we owe to the regular working of this combination during the past week; and could it be suddenly abolished, a great proportion of us would be dead before another week ended. If these most conspicuous and vital arrangements of our social structure, have arisen without the devising of any one, but through the individual efforts of citizens to satisfy their own wants; we may be tolerably certain that the less important arrangements have similarly arisen.

"But surely," it will be said, "the social changes directly

produced by law, cannot be classed as spontaneous growths. When parliaments or kings order this or that thing to be done, and appoint officials to do it, the process is clearly artificial; and society to this extent becomes a manufacture rather than a growth." No, not even these changes are exceptions, if they be real and permanent changes. The true sources of such changes lie deeper than the acts of legislators. To take first the simplest balance. We all know that the enactments of representative governments ultimately depend on the national will: they may for a time be out of harmony with it, but eventually they must conform to it. And to say that the national will finally determines them, is to say that they result from the average of individual desires; or, in other words — from the average of individual natures. A law so initiated, therefore, really grows out of the popular character.

In the case of a Government representing a dominant class, the same things holds, though not so manifestly. For the very existence of a class monopolizing all power, is due to certain sentiments in the commonalty. But for the feeling of loyalty on the part of retainers, a feudal system could not exist. We see in the protest of the Highlanders against the abolition of heritable jurisdictions, that they preferred that kind of local rule. And if to the popular nature, must thus be ascribed the growth of an irresponsible ruling class; then to the popular nature must be ascribed the social arrangements which that class creates in the pursuit of its own ends. Even where the Government is despotic, the doctrine still holds. The character of the people is, as before, the original source of this political form; and, as we have abundant proof, other forms suddenly created will not act, but rapidly retrograde to the old form. Moreover, such regulations as a despot makes, if really operative, are so because of their fitness to the social state. His acts being very much swayed by general opinion — by precedent, by the feeling of his nobles, his priesthood, his army — are in part immediate results of the national character; and when they are out of harmony with the national character, they are soon practically abrogated.

The failure of Cromwell permanently to establish a new social condition, and the rapid revival of suppressed institutions and practices after his death, show how powerless is a monarch to change the type of the society he governs. He may disturb, he may retard, or he may aid the natural process of organization; but the general coarse of this process is beyond his control. Nay, more than this is true. Those who regard the histories of societies as the histories of their great men, and think that these great men shape the fates of their societies, overlook the truth that such great men are the products of their societies. Without certain antecedents — without a certain average national character, they could neither have been generated nor could have had the culture which formed them. If their society is to some extent re-moulded by them, they were, both before and after birth, moulded by their society — were the results of all those influences which fostered the ancestral character they inherited, and gave their own early bias, their creed, morals, knowledge, aspirations. So that such social changes as are immediately traceable to individuals of unusual power, are still remotely traceable to the social causes which produced these individuals, and hence, from the highest point of view, such social changes also, are parts of the general developmental process.

Thus that which is so obviously true of the industrial structure of society, is true of its whole structure. The fact that "constitutions are not made, but grow," is simply a fragment of the much larger fact, that under all its aspects and though all its ramifications, society is a growth and not a manufacture.

A perception that there exists some analogy between the body politic and a living individual body, was early reached; and from time to time re-appeared in literature. But this perception was necessarily vague and more or less fanciful. In the absence of physiological science, and especially of those comprehensive generalizations which it has but recently reached, it was impossible to discern the real parallelisms.

The central idea of Plato's model Republic, is the correspondence between the parts of a society and the faculties of the human mind. Classifying these faculties under the heads of Reason, Will, and Passion, he classifies the members of his ideal society under what he regards as three analogous heads: — councillors, who are to exercise government; military or executive, who are to fulfil their behests; and the commonalty, bent on gain and selfish gratification. In other words, the ruler, the warrior, and the craftsman, are, according to him, the analogues of our reflective, volitional, and emotional powers. Now even were there truth in the implied assumption of a parallelism between the structure of a society and that of a man, this classification would be indefensible. It might more truly be contended that, as the military power obeys the commands of the Government, it is the Government which answers to the Will; while the military power is simply an agency set in motion by it. Or, again, it might be contended that whereas the Will is a product of predominant desires, to which the Reason serves merely as an eye, it is the craftsmen, who, according to the alleged analogy, ought to be the moving power of the warriors.

Hobbes sought to establish a still more definite parallelism: not, however between a society and the human mind, but between a society and the human body. In the introduction to the work in which he developes this conception, he says —

"For by art is created that great Leviathan called a Commonwealth, or State, in Latin Civitas, which is but an artificial man; though of greater stature and strength than the natural, for whose protection and defence it was intended, and in which the *sovereignty* is an artificial *soul,* as giving life and motion to the whole body; the *magistrates* and other *officers* of judicature and execution, artificial *joints; reward* and *punishment,* by which, fastened to the seat of the sovereignty, every joint and member is moved to perform his duty, are the *nerves,* that do the same in the body natural; the *wealth* and *riches* of all the particular members are the *strength; salus populi,* the *people's safety,* its *business; counsellors,* by whom all things needful for it to know are suggested unto it, are the *memory; equity* and *laws* an artificial *reason* and *will; concord, health; sedition, sickness; civil war, death."*

And Hobbes carries this comparison so far as actually to give a drawing of the Leviathan — a vast human-shaped figure, whose body and limbs are made up of multitudes of men. Just noting that these different analogies asserted by Plato and Hobbes, serve to cancel each other (being, as they are, so completely at variance), we may say that on the whole those of Hobbes are the more plausible. But they are full of inconsistencies. If the sovereignty is the *soul* of the body politic, how can it be that magistrates, who are a kind of deputy-sovereigns, should be comparable to *joints?* Or, again, how can the three mental functions, memory, reason, and will, be severally analogous, the first to counsellors, who are a class of public officers, and the other two to equity and laws, which are not classes of officers, but abstractions? Or, once more, if magistrates are the artificial joints of society, how can reward and punishment be its nerves? Its nerves must surely be some class of persons. Reward and punishment must in societies, as in individuals, be *conditions* of the nerves, and not the nerves themselves.

But the chief errors of these comparisons made by Plato and Hobbes, lie much deeper. Both thinkers assume that the organization of a society is comparable, not simply to the organization of a living body in general, but to the organization of the human body in particular. There is no warrant whatever for assuming this. It is in no way implied by the evidence; and is simply one of those fancies which we commonly find mixed up with the truths of early speculation. Still more erroneous are the two conceptions in this, that they construe a society as an artificial structure. Plato's model republic — his ideal of a healthful body politic — is to be consciously put together by men; just as a watch might be: and Plato manifestly thinks of societies in general as thus originated. Quite specifically does Hobbes express this view. "For by *art,*" be says, "is created that great Leviathan called a Commonwealth." And he even goes so far as to compare the supposed social contract, from which a society suddenly originates, to the creation of a man by the divine fiat. Thus they both fall into the extreme inconsistency of considering a community as similar in structure to a human being, and yet as produced in the same way as an artificial mechanism — in nature, an organism; in history, a machine.

Notwithstanding errors, however, these speculations have considerable significance. That such analogies, crudely as they are thought out, should have been alleged by Plato and Hobbes and many others, is a reason for suspecting that *some* analogy exists. The untenableness of the particular comparisons above instanced, is no ground for denying an essential parallelism; for early ideas are usually but vague adumbrations of the truth. Lacking the great generalizations of biology, it was, as we have said, impossible to trace out the real relations of social organizations to organizations of another order. We propose here to show what are the analogies which modern science discloses to us.

Let us set out by succinctly stating the points of similarity and the points of difference. Societies agree with individual organisms in four conspicuous peculiarities: —

1. That commencing as small aggregations, they insensibly augment in mass: some of them eventually reaching ten thousand times what they originally were.

2. That while at first so simple in structure as to be considered structureless, they assume, in the course of their growth, a continually-increasing complexity of structure.

3. That though in their early, undeveloped states, there exists in them scarcely any mutual dependence of parts, their parts gradually acquire a mutual dependence; which becomes at last so great, that the activity and life of each part is made possible only by the activity and life of the rest.

4. That the life and development of a society is independent of, and far more prolonged than, the life and development of any of its component units; who are severally born, grow, work, reproduce, and die, while the body politic composed of them survives generation after generation, increasing in mass, completeness of structure, and functional activity.

These four parallelisms will appear the more significant the more we contemplate them. While the points specified, are points in which societies agree with individual organisms, they are points in which individual organisms agree with each other, and disagree with all things else. In the course of its existence, every plant and animal increases in mass, in a way not paralleled by inorganic objects: even such inorganic objects as crystals, which arise by growth, show us no such definite relation between growth and existence as organisms do. The orderly progress from simplicity to complexity, displayed by bodies politic in common with all living bodies, is a characteristic which distinguishes living bodies from the inanimate bodies amid which they move. That functional dependence of parts, which is scarcely more manifest in animals or plants than nations, has no counterpart elsewhere. And in no aggregate except an organic, or a social one, is there a perpetual removal and replacement of parts, joined with a continued integrity of the whole.

Moreover, societies and organisms are not only alike in these peculiarities, in which they are unlike all other things; but the highest societies, like the highest organisms, exhibit them in the greatest degree. We see that the lowest animals do not increase to anything like the sizes of the higher ones; and, similarly, we see that aboriginal societies are comparatively limited in their growths. In complexity, our large civilized nations as much exceed primitive savage tribes, as a vertebrate animal does a zoophyte. Simple communities, like simple creatures, have so little mutual dependence of parts, that subdivision or mutilation causes but little inconvenience; but from complex communities, as from complex creatures, you cannot remove any considerable organ without producing great disturbance or death of the rest. And in societies of low type, as in inferior animals, the life of the aggregate, often cut short by division or dissolution, exceeds in length the lives of the component units, very far less than in civilized communities and superior animals; which outlive many generations of their component units.

On the other hand, the leading differences between societies and individual organisms are these: —

1. That societies have no specific external forms. This, however, is a point of contrast which loses much of its importance, when we remember that throughout the vegetal kingdom, as well as in some lower divisions of the animal kingdom, the forms are often very indefinite — definiteness being rather the exception than the rule; and that they are manifestly in part determined by surrounding physical circumstances, as the forms of societies are. If, too, it should eventually be shown, as we believe it will, that the form of every species of organism has resulted from the average play of the external forces to which it has been subject during its evolution as a species; then, that the external forms of societies should depend, as they do, on surrounding conditions, will be a further point of community.

2. That though the living tissue whereof an individual organism consists, forms a continuous mass, the living elements of a society do not form a continuous mass; but are more or less widely dispersed over some portion of the Earth's surface. This, which at first sight appears to be a fundamental distinction, is one which yet to a great extent disappears when we contemplate all the facts. For, in the lower divisions of the animal and vegetal kingdoms, there are types of organization much more nearly allied, in this respect, to the organization of a society, than might be supposed — types in which the living units essentially composing the mass, are dispersed through an inert substance, that

can scarcely be called living in the full sense of the word. It is thus with some of the *Protococci* and with the *Nostoceœ,* which exist as cells imbedded in a viscid matter. It is so, too, with the *Thalassicollœ* — bodies that are made up of differentiated parts, dispersed through an undifferentiated jelly. And throughout considerable portions of their bodies, some of the *Acalephœ* exhibit more or less distinctly this type of structure.

Indeed, it may be contended that this is the primitive form of all organization; seeing that, even in the highest creatures, as in ourselves, every tissue developes out of what physiologists call a blastema — an unorganized though organizable substance, through which organic points are distributed. Now this is very much the case with a society. For we must remember that though the men who make up a society, are physically separate and even scattered; yet that the surface over which they are scattered is not one devoid of life, but is covered by life of a lower order which ministers to their life. The vegetation which clothes a country, makes possible the animal life in that country; and only through its animal and vegetal products can such a country support a human society. Hence the members of the body politic are not to be regarded as separated by intervals of dead space; but as diffused through a space occupied by life of a lower order. In our conception of a social organism, we must include all that lower organic existence on which human existence, and therefore social existence, depends. And when we do this, we see that the citizens who make up a community, may be considered as highly vitalized units surrounded by substances of lower vitality, from which they draw their nutriment: much as in the cases above instanced. Thus, when examined, this apparent distinction in great part disappears.

3. That while the ultimate living elements of an individual organism, are mostly fixed in their relative positions, those of the social organism are capable of moving from place to place, seems a marked disagreement. But here, too, the disagreement is much less than would be supposed. For while citizens are locomotive in their private capacities, they are fixed in their public capacities. As farmers, manufacturers, or traders, men carry on their business at the same spots, often throughout their whole lives; and if they go away occasionally, they leave behind others to discharge their functions in their absence. Each great centre of

production, each manufacturing town or district, continues always in the same place; and many of the firms in such town or district, are for generations carried on either by the descendants or successors of those who founded them. Just as in a living body, the cells that make up some important organ, severally perform their functions for a time and then disappear, leaving others to supply their places; so, in each part of a society, the organ remains, though the persons who compose it change. Thus, in social life, as in the life of an animal, the units as well as the larger agencies formed of them, are in the main stationary as respects the places where they discharge their duties and obtain their sustenance. And hence the power of individual locomotion does not practically affect the analogy.

4. The last and perhaps the most important distinction, is, that while in the body of an animal, only a special tissue is endowed with feeling; in a society, all the members are endowed with feeling. Even this distinction, however, is by no means a complete one. For in some of the lowest animals, characterized by the absence of a nervous system, such sensitiveness as exists is possessed by all parts. It is only in the more organized forms that feeling is monopolized by one class of the vital elements. Moreover, we must remember that societies, too, are not without a certain differentiation of this kind. Though the units of a community are all sensitive, yet they are so in unequal degrees. The classes engaged in agriculture and laborious occupations in general, are much less susceptible, intellectually and emotionally, than the rest; and especially less so than the classes of highest mental culture. Still, we have here a tolerably decided contrast between bodies politic and individual bodies. And it is one which we should keep constantly in view. For it reminds us that while in individual bodies, the welfare of all other parts is rightly subservient to the welfare of the nervous system, whose pleasurable or painful activities make up the good or evil of life; in bodies politic, the same thing does not hold, or holds to but a very slight extent. It is well that the lives of all parts of an animal should be merged in the life of the whole; because the whole has a corporate consciousness capable of happiness or misery. But it is not so with a society; since its living units do not and cannot lose individual consciousness; and since the community as a whole has no corporate consciousness. And this is an everlasting

reason why the welfare of citizens cannot rightly be sacrificed to some supposed benefit of the State; but why, on the other hand, the State is to be maintained solely for the benefit of citizens. The corporate life must here be subservient to the lives of the parts; instead of the lives of the parts being subservient to the corporate life.

Such, then, are the points of analogy and the points of difference. May we not say that the points of difference serve but to bring into clearer light the points of analogy. While comparison makes definite the obvious contrasts between organisms commonly so called, and the social organism; it shows that even these contrasts are not so decided as was to be expected. The indefiniteness of form, the discontinuity of the parts, the mobility of the parts, and the universal sensitiveness, are not only peculiarities of the social organism which have to be stated with considerable qualifications; but they are peculiarities to which the inferior classes of animals present approximations. Thus we find but little to conflict with the all-important analogies. That societies slowly augment in mass; that they progress in complexity of structure; that at the same time their parts become more mutually dependent; that their living units are removed and replaced without destroying their integrity; and further, that the extents to which they display these peculiarities are proportionate to their vital activities; are traits that societies have in common with organic bodies. And these traits in which they agree with organic bodies and disagree with all other things — these traits which in truth specially characterize organic bodies, entirely subordinate the minor distinctions: such distinctions being scarcely greater than those which separate one half of the organic kingdom from the other. The *principles* of organization are the same; and the differences are simply differences of application.

Here ending this general survey of the facts which justify the comparison of a society to a living body; let us look at them in detail. We shall find that the parallelism becomes the more marked the more closely it is traced.

The lowest animal and vegetal forms — *Protozoa* and *Protophyta* — are chiefly inhabitants of the water. They are minute bodies, most of which are made individually visible only by the microscope. All of them are extremely simple in structure; and some of them, as the *Rhizopods,* almost structureless. Multiplying, as they ordinarily do, by the spontaneous division of their bodies, they produce halves, which may either become quite separate and move away in different directions, or may continue attached. By the repetition of this process of fission, aggregations of various sizes and kinds are formed. Among the *Protophyta* we have some classes, as the *Diatomaceæ* and the Yeast-plant, in which the individuals may be either separate, or attached in groups of two, three, four, or more; other classes in which a considerable number of individual cells are united into a thread (*Conferva, Monilia);* others in which they form a net work *(Hydrodictyon);* others in which they form plates (*Ulva);* and others in which they form masses *(Laminaria, Agaricus):* all which vegetal forms, having no distinction of root, stem, or leaf, are called *Thallogens.* Among the *Protozoa* we find parallel facts. Immense numbers of *Amœba*-like creatures, massed together in a framework of horny fibres, constitute Sponge. In the *Foraminifera,* we see smaller groups of such creatures arranged into more definite shapes. Not only do these almost structureless *Protozoa* unite into regular or irregular aggregations of various sizes; but among some of the more organized ones, as the *Vorticellæ,* there are also produced clusters of individuals, proceeding from a common stock. But these little societies of monads, or cells, or whatever else we may call them, are societies only in the lowest sense: there is no subordination of parts among them — no organization. Each of the component units lives by and for itself; neither giving nor receiving aid. There is no mutual dependence, save that consequent on mere mechanical union.

Now do we not here discern analogies to the first stages of human societies? Among the lowest races, as the Bushmen, we find but incipient aggregation: sometimes single families; sometimes two or three families wandering about together. The number of associated units is small and variable; and their union inconstant. No division of labour exists except between the sexes; and the only kind of mutual aid is that of joint attack or defence. We see nothing beyond an undifferentiated group of individuals, forming the germ of a society; just as in the homogeneous groups of cells above described, we see only the initial stage of animal and vegetal organization.

The comparison may now be carried a step higher. In the vegetal kingdom we pass from the *Thallogens,* consisting of mere masses of similar cells, to the *Acrogens,* in which the cells are not similar throughout the whole mass; but are here aggregated into a structure serving as leaf, and there into a structure serving as root: thus forming a whole in which there is a certain subdivision of functions among the units; and therefore a certain mutual dependence. In the animal kingdom we find analogous progress. From mere unorganized groups of cells, or cell-like bodies, we ascend to groups of such cells arranged into parts that have different duties. The common Polype, from whose substance may be separated individual cells which exhibit, when detached, appearances and movements like those of the solitary *Amœba,* illustrates this stage. The component units, though still showing great community of character, assume somewhat diverse functions in the skin, in the internal surface, and in the tentacles. There is a certain amount of "physiological division of labour."

Turning to societies, we find these stages paralleled in the majority of aboriginal tribes. When, instead of such small variable groups as are formed by Bushmen, we come to the larger and more permanent groups formed by savages not quite so low, we begin to find traces of social structure. Though industrial organization scarcely shows itself, except in the different occupations of the sexes; yet there is always more or less of governmental organization. While all the men are warriors and hunters, only a part of them are included in the council of chiefs; and in this council of chiefs some one has commonly supreme authority. There is thus a certain distinction of classes and powers; and through this slight specialization of functions, is effected a rude co-operation among the increasing mass of individuals, whenever the society has to act in its corporate capacity. Beyond this analogy in the slight extent to which organization is carried, there is analogy in the indefiniteness of the organization. In the *Hydra,* the respective parts of the creature's substance have many functions in common. They are all contractile; omitting the tentacles, the whole of the external surface can give origin to young *hydrœ;* and when turned inside out, stomach performs the duties of skin, and skin the duties of stomach. In aboriginal societies such differentiations as exist are similarly imperfect. Notwithstanding distinctions of rank, all persons maintain themselves by their own exertions. Not only do the head men of the tribe, in common with the rest, build their own huts, make their own weapons, kill their own food; but the chief does the like. Moreover, in the rudest of these tribes, such governmental organization as exists is very inconstant. It is frequently changed by violence or treachery, and the function of ruling assumed by other members of the community. Thus between the rudest societies and some of the lowest forms of animal life there is analogy alike in the slight extent to which organization is carried, in the indefiniteness of this organization, and in its want of fixity.

A further complication of the analogy is at hand. From the aggregation of units into organized groups, we pass to the multiplication of such groups, and their coalescence into compound groups. The *Hydra,* when it has reached a certain bulk, puts forth from its surface a bud, which, growing and gradually assuming the form of the parent, finally becomes detached; and by this process of gemmation, the creature peoples the adjacent water with others like itself. A parallel process is seen in the multiplication of those lowly-organized tribes above described. One of them having increased to a size that is either too great for coordination under so rude a structure, or else that is greater than the surrounding country can supply with game and other wild food, there arises a tendency to divide; and as in such communities there are ever occurring quarrels, jealousies, and other causes of division, there soon comes an occasion on which a part of the tribe separates under the leadership of some subordinate chief, and migrates. This process being from time to time repeated, an extensive region is at length occupied with numerous separate tribes descended from a common ancestry. The analogy by no means ends here. Though in the common *Hydra,* the young ones that bud out from the parent soon become detached and independent; yet throughout the rest of the class *Hydrozoa,* to which this creature belongs, the like does not generally happen. The successive individuals thus developed continue attached; give origin to other such individuals which also continue attached; and so there results a compound animal. As in the *Hydra* itself, we find an aggregation of units which, considered separately, are akin to the lowest *Protozoa;* so here, in a *Zoophyte,* we find an aggregation of such aggregations. The like is also seen throughout the extensive family of *Polyzoa* or *Molluscoida.* The Ascidian Mollusks, too, in their many varied forms, show us the same thing: exhibiting, at the same time, various degrees of union subsisting among the component individuals. For while in the *Salpœ* the component individuals adhere so slightly that a blow on the vessel of water in which they are floating will separate them; in the *Botryllidœ* there exists a vascular connexion between them, and a common circulation.

Now in these various forms and degrees of aggregation, may we not see paralleled the union of groups of connate tribes into nations? Though in regions where circumstances permit, the separate tribes descended from some original tribe, migrate in all directions, and become far removed and quite separate; yet, in other cases, where the territory presents barriers to distant migration, this does not happen: the small kindred communities are held in closer contact, and eventually become more or less united into a nation. The contrast between the tribes of American Indians and the Scottish clans, illustrates this. And a glance at our own early history, or the early histories of continental nations, shows this fusion of small simple communities taking place in various ways and to various extents. As says M. Guizot, in his history of "The Origin of Representative Government," —

"By degrees, in the midst of the chaos of the rising society, small aggregations are formed which feel the want of alliance and union with each other.... Soon inequality of strength is displayed among neighbouring aggregations. The strong tend to subjugate the weak, and usurp at first the rights of taxation and military service. Thus political authority leaves the aggregations which first instituted it, to take a wider range."

That is to say, the small tribes, clans, or feudal unions, sprung mostly from a common stock, and long held in contact as occupants of adjacent lands, gradually get united in other ways than by mere adhesion of race and proximity.

A further series of changes begins now to take place; to which, as before, we shall find analogies in individual organisms. Returning again to the *Hydrozoa,* we observe that in the simplest of the compound forms, the connected individuals developed from a common stock, are alike in structure, and perform like functions: with the exception, indeed, that here and there a bud, instead of developing into a stomach, mouth, and tentacles, becomes an egg-sac. But with the oceanic *Hydrozoa,* this is by no means the case. In the *Calycophoridœ,* some of the polypes growing from the common germ, become developed and modified into large, long, sack-like bodies, which by their rhythmical contractions move through the water, dragging the community of polypes after them. In the *Physophoridæ,* a variety of organs similarly arise by transformation of the budding polypes; so that in creatures like the *Physalia,* commonly known as the "Portuguese Man-of-war," instead of that tree-like group of similar individuals forming the original type of the class, we have a complex mass of unlike parts fulfilling unlike duties. As an individual *Hydra* may be regarded as a group of *Protozoa,* which have become partially metamorphosed into different organs; so a *Physalia* is, morphologically considered, a group of *Hydræ* of which the individuals have been variously transformed to fit them for various functions.

This differentiation upon differentiation, is just what takes place in the evolution of a civilized society. We observed how, in the small communities first formed, there arises a certain simple political organization — there is a partial separation of classes having different duties. And now we have to observe how, in a nation formed by the fusion of such small communities, the several sections, at first alike in structures and modes of activity, gradually become unlike in both — gradually become mutually-dependent parts, diverse in their natures and functions.

The doctrine of the progressive division of labour, to which we are here introduced, is familiar to all readers. And further, the analogy between the economical division of labour and the "physiological division of labour," is so striking, as long since to have drawn the attention of scientific naturalists: so striking, indeed, that the expression "physiological division of labour," has been suggested by it. It is not needful, therefore, that we should treat this part of our subject in great detail. We shall content ourselves with noting a few general and significant facts, not manifest on a first inspection.

Throughout the whole animal kingdom, from the *Cœlenterata* upwards, the first stage of evolution is the same. Equally in the germ of a polype and in the human ovum, the aggregated mass of cells out of which the creature is to arise, gives origin to a peripheral layer of cells, slightly differing from the rest which they include; and this layer subsequently divides into two — the inner, lying in contact with the included yelk, being called the mucous layer, and the outer, exposed to surrounding agencies, being called the serous layer: or, in the terms used by Prof. Huxley, in describing the development of the *Hydrozoa* — the endoderm and ectoderm. This primary division marks out a fundamental contrast of parts in the future organism. From the mucous layer, or endoderm, is developed the apparatus of nutrition; while from the serous layer, or ectoderm, is developed the apparatus of external action. Out of the one arise the organs by which food is prepared and absorbed, oxygen imbibed, and blood purified; while out of the other arise the nervous, muscular, and osseous systems, by whose combined actions the movements of the body as a whole are effected. Though this is not a rigorously-correct distinction, seeing that some organs involve both of these primitive membranes, yet high authorities agree in stating it as a broad general distinction.

Well, in the evolution of a society, we see a primary differentiation of analogous kind; which similarly underlies the whole future structure. As already pointed out, the only manifest contrast of parts in primitive societies, is that between the governing and the governed. In the least organized tribes, the council of chiefs may be a body of men distinguished simply by greater courage or experience. In more organized tribes, the chief-class is definitely separated from the lower class, and often regarded as different in nature — sometimes as god-descended. And later, we find these two becoming respectively freemen and slaves, or nobles and serfs. A glance at their respective functions, makes it obvious that the great divisions thus early formed, stand to each other in a relation similar to that in which the primary divisions of the embryo stand to each other. For, from its first appearance, the class of chiefs is that by which the external acts of the society are controlled: alike in war, in negotiation, and in migration. Afterwards, while the upper class grows distinct from the lower, and at the same time becomes more and more exclusively regulative and defensive in its functions, alike in the persons of kings and subordinate rulers, priests, and military leaders; the inferior class becomes more and more exclusively occupied in providing the necessaries of life for the community at large. From the soil, with which it comes in most direct contact, the mass of the people takes up and prepares for use, the food and such rude articles of manufacture as are known; while the overlying mass of superior men, maintained by the working population, deals with circumstances external to the community— circumstances with which, by position, it is more immediately concerned. Ceasing by-and-by to have any knowledge of, or power over, the concerns of the society as a whole, the serf-class becomes devoted to the processes of alimentation; while the noble class, ceasing to take any part in the processes of alimentation, becomes devoted to the co-ordinated movements of the entire body politic.

Equally remarkable is a further analogy of like kind. After the mucous and serous layers of the embryo have separated, there presently arises between the two, a third, known to physiologists as the vascular layer — a layer out of which are developed the chief ·blood-vessels. The mucous layer absorbs nutriment from the mass of yelk it encloses; this nutriment has to be transferred to the overlying serous layer, out of which the nervo-muscular system is being developed; and between the two arises a vascular system by which the transfer is effected — a system of vessels which continues ever after to be the transferrer of nutriment from the places where it is absorbed and prepared, to the places where it is needed for growth and repair. Well, may we not trace a parallel step in social progress?

Between the governing and the governed, there at first exists no intermediate class; and even in some societies that have reached considerable sizes, there are scarcely any but the nobles and their kindred on the one baud, and the serfs on the other: the social structure being such, that the transfer of commodities takes place directly from slaves to their masters. But in societies of a higher type, there grows up between these two primitive classes, another — the trading or middle class. Equally, at first as now, we may see that, speaking generally, this middle class is the analogue of the middle layer in the embryo. For all traders are essentially distributors. Whether they be wholesale dealers, who collect into large masses the commodities of various producers; or whether they be retailers, who divide out to those who want them, the masses of commodities thus collected together; all mercantile men are agents of transfer from the places where things are produced to the places where they are consumed. Thus the distributing apparatus of a society, answers to the distributing apparatus of a living body; not only in its functions, but in its intermediate origin and subsequent position, and in the time of its appearance.

Without enumerating the minor differentiations which these three great classes afterwards undergo, we will merely note that throughout, they follow the same general law with the differentiations of an individual organism. In a society, as in a rudimentary animal, we have seen that the most general and broadly contrasted divisions are the first to make their appearance; and of the subdivisions it continues true in both cases, that they arise in the order of decreasing generality.

Let us observe next, that in the one case as in the other, the specializations are at first very incomplete; and become more complete as organization progresses. We saw that in primitive tribes, as in the simplest animals, there remains much community of function between the parts that are nominally different — that, for instance, the class of chiefs long remain industrially the same as the inferior class; just as in a *Hydra,* the property of contractility is possessed by the units of the endoderm as well as by those of the ectoderm. We noted also how, as the society advanced, the two great primitive classes partook less and less of each other's functions. And we have here to remark, that all subsequent specializations are at first vague, and gradually become distinct. "In the infancy of society," says M. Guizot, "everything is confused and uncertain; there is as yet no fixed and precise line of demarcation between the different powers in a state." "Originally kings lived like other landowners, on the incomes derived from their own private estates." Nobles were petty kings; and kings only the most powerful nobles. Bishops were feudal lords and military leaders. The right of coming money was possessed by powerful subjects, and by the Church, as well as by the king. Every leading man exercised alike the functions of landowner, farmer, soldier, statesman, judge. Retainers were now soldiers, and now labourers, as the day required. But by degrees the Church has lost all civil jurisdiction; the State has exercised less and less control over religious teaching; the military class has grown a distinct one; handicrafts have concentrated in towns; and the spinning-wheels of scattered farmhouses, have disappeared before the machinery of manufacturing districts. Not only is all progress from the homogeneous to the heterogeneous; but at the same time it is from the indefinite to the definite.

Another fact which should not be passed over, is that in the evolution of a large society out of an aggregation of small ones, there is a gradual obliteration of the original lines of separation — a change to which, also, we may see analogies in living bodies. Throughout the sub-kingdom *Annulosa,* this is clearly and variously illustrated. Among the lower types of this sub-kingdom, the body consists of numerous segments that are alike in nearly every particular. Each has its external ring; its pair of legs, if the creature has legs; its equal portion of intestines, or else its separate stomach; its equal portion of the great blood-vessel, or, in some cases, its separate heart; its equal portion of the nervous cord, and, perhaps, its separate pair of ganglia. But in the highest types, as in the large *Crustacea,* many of the segments are completely fused together; and the internal organs are no longer uniformly repeated in all the segments. Now the segments of which nations at first consist, lose their separate external and internal structures in a similar manner. In feudal times, the minor communities governed by feudal lords, were severally organized in the same rude way; and were held together only by the fealty of their respective rulers to some suzerain. But along with the growth of a central power, the demarcations of these local communities disappeared; and their separate organizations merged into the general organization. The like is seen on a larger scale in the fusion of England, Whales, Scotland, and Ireland; and, on the Continent, in the coalescence of provinces into kingdoms. Even in the disappearance of law-made divisions, the process is analogous. Among the Anglo-Saxons, England was divided into tithings, hundreds, and counties: there were county courts, courts of hundred, and courts of tithing. The courts of tithing disappeared first; then the courts of hundred, which have, however, left traces; while the county-jurisdiction still exists.

But chiefly it is to be noted, that there eventually grows up an organization which has no reference to these original divisions, but traverses them in various directions, as is the case in creatures belonging to the sub-kingdom just named; and, further, that in both cases it is the sustaining organization which thus traverses old boundaries, while in both cases it is the governmental, or co-ordinating organization in which the original boundaries continue traceable. Thus, in the highest *Annulosa,* the exo-skeleton and the muscular system, never lose all traces of their primitive segmentation; but throughout a great part of the body, the contained viscera do not in the least conform to the external divisions. Similarly, with a nation, we see that while, for governmental purposes, such divisions as counties and parishes still exist, the structure developed for carrying on the nutrition of society, wholly ignores these boundaries: our great cotton-manufacture spreads out of Lancashire into North Derbyshire; Leicestershire and Nottinghamshire have long divided the stocking-trade between them; one great centre for the production of iron and iron-goods, includes parts of Warwickshire, Staffordshire, Worcestershire; and those various specializations of agriculture which have made different parts of England noted for different products, show no more respect to county-boundaries than do our growing towns to the boundaries of parishes.

If, after contemplating these analogies of structure, we inquire whether there are any such analogies between the processes of organic change, the answer is — yes. The causes which lead to increase of bulk in any part of the body politic, are of like nature with those which lead to increase of bulk in any part of an individual body. In both cases the antecedent is greater functional activity, consequent on greater demand. Each limb, viscus, gland, or other member of an animal, is developed by exercise — by actively discharging the duties which the body at large requires of it; and similarly, any class of labourers or artisans, any manufacturing centre, or any official agency, begins to enlarge when the community devolves on it an increase of work. In each case, too, growth has its conditions and its limits. That any organ in a living being may grow by exercise, there needs a due supply of blood: all action implies waste; blood brings the materials for repair; and before there can be growth, the quantity of blood supplied must be more than that requisite for repair.

So is it in a society. If to some district which elaborates for the community particular commodities — say the woollens of Yorkshire — there comes an augmented demand; and if, in fulfilment of this demand, a certain expenditure and wear of the manufacturing organization are incurred; and if, in payment for the extra supply of woollens sent away, there comes back only such quantity of commodities as replaces the expenditure, and makes good the waste of life and machinery; there can clearly be no growth. That there may be growth, the commodities obtained in return must be more than sufficient for these ends; and just in proportion as the surplus is great will the growth be rapid. Whence it is manifest that what in commercial affairs we call *profit,* answers to the excess of nutrition over waste in a living body. Moreover, in both cases, when the functional activity is high and the nutrition defective, there results not growth but decay. If in an animal, any organ is worked so hard that the channels which bring blood cannot furnish enough for repair, the organ dwindles; and if in the body politic, some part has been stimulated into great productivity, and cannot afterwards get paid for all its produce, certain of its members become bankrupt, and it decreases in size.

One more parallelism to be here noted, is, that the different parts of the social organism, like the different parts of an individual organism, compete for nutriment; and severally obtain more or less of it according as they are discharging more or less duty. If a man's brain be overexcited, it will abstract blood from his viscera and stop digestion; or digestion actively going on, will so affect the circulation through the brain as to cause drowsiness; or great muscular exertion will determine such a quantity of blood to the limbs, as to arrest digestion or cerebral action, as the case may be. So, likewise, in a society, it frequently happens that great activity in some one direction, causes partial arrests of activity elsewhere, by abstracting capital, that is commodities: as instance the way in which the sudden development of our railway-system hampered commercial operations; or the way in which the raising of a large military force temporarily stops the growth of leading industries.

The last few paragraphs introduce the next division of our subject. Almost unawares we have come upon the analogy which exists between the blood of a living body, and the circulating mass of commodities in the body politic. We have now to trace out this analogy from its simplest to its most complex manifestations.

In the lowest animals there exists no blood properly so called. Through the small aggregation of cells which make up a *Hydra,* permeate the juices absorbed from the food. There is no apparatus for elaborating a concentrated and purified nutriment, and distributing it among the component units; but these component units directly imbibe the unprepared nutriment, either from the digestive cavity or from each other. May we not say that this is what takes place in an aboriginal tribe? All its members severally obtain for themselves the necessaries of life in their crude states; and severally prepare them for their own uses as well as they can. When there arises a decided differentiation between the governing and the governed, some amount of transfer begins between those inferior individuals, who, as workers, come directly in contact with the products of the earth, and those superior ones who exercise the higher functions — a transfer parallel to that which accompanies the differentiation of the ectoderm from the endoderm. In the one case, as in the other, however, it is a transfer of products that are little if at all prepared; and takes place directly from the unit which obtains to the unit which consumes, without entering into any general current.

Passing to larger organisms — individual and social — we find the first advance upon this arrangement. Where, as among the compound *Hydrozoa,* there is an aggregation of many such primitive groups as form *Hydrœ;* or where, as in a *Medusa,* one of these groups has become of great size; there exist rude channels running throughout the substance of the body: not however, channels for the conveyance of prepared nutriment, but mere prolongations of the digestive cavity, through which the crude chyle-aqueous fluid reaches the remoter parts, and is moved backwards and forwards by the creature's contractions. Do we not find in some of the more advanced primitive communities, an analogous condition? When the men, partially or fully united into one society, become numerous — when, as usually happens, they cover a surface of country not everywhere alike in its products — when, more especially, there arise considerable classes that are not industrial; some process of exchange and distribution inevitably arises. Traversing here and there the earth's surface, covered by that vegetation on which human life depends, and in which, as we say, the units of a society are imbedded, there are formed indefinite paths, along which some of the necessaries of life occasionally pass, to be bartered, for others which presently come back along the same channels. Note, however, that at first little else but crude commodities are thus transferred — fruits, fish, pigs or cattle, skins, etc.: there are few, if any, manufactured products or articles prepared for consumption. And note further, that such distribution of these unprepared necessaries of life as takes place, is but occasional — goes on with a certain slow, irregular rhythm.

Further progress in the elaboration and distribution of nutriment, or of commodities, is a necessary accompaniment of further differentiation of functions in the individual body or in the body politic. As fast as each organ of a living animal becomes confined to a special action, it must become dependent on the rest for all those materials which its position and duty do not permit it to obtain for itself; in the same way that, as fast as each particular class of a community becomes exclusively occupied in producing its own commodity, it must become dependent on the rest for the other commodities it needs. And, simultaneously, a more perfectly-elaborated blood will result from a highly-specialized group of nutritive organs, severally adapted to prepare its different elements; in the same way that the stream of commodities circulating throughout a society, will be of superior quality in proportion to the greater division of labour among the workers. Observe, also, that in either case the circulating mass of nutritive materials, besides coming gradually to consist of better ingredients, also grows more complex. An increase in the number of the unlike organs which add to the blood their waste matters, and demand from it the different materials they severally need, implies a blood more heterogeneous in composition — an *à priori* conclusion which, according to Dr. Williams, is inductively confirmed by examination of the blood throughout the various grades of the animal kingdom. And similarly, it is manifest that as fast as the division of labour among the classes of a community, becomes greater, there must be an increasing heterogeneity in the currents of merchandise flowing throughout that community.

The circulating mass of nutritive materials in individual organisms and in social organisms, becoming alike better in the quality of its ingredients and more heterogeneous in composition, as the type of structure becomes higher; eventually has added to it in both cases another element, which is not itself nutritive, but facilitates the process of nutrition. We refer, in the case of the individual organism, to the blood-discs; and in the case of the social organism, to money. This analogy has been observed by Liebig, who in his "Familiar Letters on Chemistry," says:

"Silver and gold have to perform in the organization of the

State, the same function as the blood corpuscles in the human organization. As these round discs, without themselves taking an immediate share in the nutritive process, are the medium, the essential condition of the change of matter, of the production of the heat, and of the force by which the temperature of the body is kept up and the motions of the blood and all the juices are determined, so has gold become the medium of all activity in the life of the State."

And blood-corpuscles being like money in their functions, and in the fact that they are not consumed in nutrition, he further points out, that the number of them which in a considerable interval flows through the great centres, is enormous when compared with their absolute number; just as the quantity of money which annually passes through the great mercantile centres, is enormous when compared with the total quantity of money in the kingdom. Nor is this all. Liebig has omitted the significant circumstance, that only at a certain stage of organization does this element of the circulation make its appearance. Throughout extensive divisions of the lower animals, the blood contains no corpuscles; and in societies of low civilization, there is no money.

Thus far, we have considered the analogy between the blood in a living body and the consumable and circulating commodities in the body politic. Let us now compare the appliances by which they are respectively distributed. We shall find in the development of these appliances, parallelisms not less remarkable than those above set forth. Already we have shown that, as classes, wholesale and retail distributors discharge in a society, the office which the vascular system discharges in an individual creature; that they come into existence later than the other two great classes, as the vascular layer appears later than the mucous and serous layers; and that they occupy a like intermediate position. Here, however, it remains to be pointed out that a complete conception of the circulating system in a society, includes not only the active human agents who propel the currents of commodities, and regulate their distribution; but includes, also, the channels of communication. It is the formation and arrangement of these, to which we now direct attention.

Going back once more to those lower animals in which there is found nothing but a partial diffusion, not of blood, but only of crude nutritive fluids, it is to be remarked that the channels through which the diffusion takes place, are mere excavations through the half-organized substance of the body: they have no lining membranes, but are mere *lacunæ* traversing a rude tissue. Now countries in which civilization is but commencing, display a like condition: there are no roads properly so called; but the wilderness of vegetal life covering the earth's surface, is pierced by tracks, through which the distribution of crude commodities takes place. And while in both cases, the acts of distribution occur only at long intervals (the currents, after a pause, now setting towards a general centre, and now away from it), the transfer is in both cases slow and difficult. But among other accompaniments of progress, common to animals and societies, comes the formation of more definite and complete channels of communication. Blood-vessels acquire distinct walls; roads are fenced and gravelled. This advance is first seen in those roads or vessels that are nearest to the chief centres of distribution; while the peripheral roads and peripheral vessels, long continue in their primitive states. At a yet later stage of development, where comparative finish of structure is found throughout the system as well as near the chief centres, there remains in both cases the difference, that the main channels are comparatively broad and straight, while the subordinate ones are narrow and tortuous in proportion to their remoteness.

Lastly, it is to be remarked that there ultimately arise in the higher social organisms, as in the higher individual organisms, main channels of distribution still more distinguished by their perfect structures, their comparative straightness, and the absence of those small branches which the minor channels perpetually give off. And in railways we also see, for the first time in the social organism, a specialization with respect to the directions of the currents — a system of double channels conveying currents in opposite directions, as do the arteries and veins of a well-developed animal.

These parallelisms in the evolutions and structures of the circulating systems, introduce us to others in the kinds and rates of the movements going on through them. In the lowest societies, as in the lowest creatures, the distribution of crude nutriment is by slow gurgitations and re-gurgitations. In creatures that have rude vascular systems, as in societies that are beginning to have roads and some transfer of commodities along them, there is no regular circulation in definite courses; but instead, periodical changes of the currents — now towards this point, and now towards that. Through each part of an inferior mollusk's body, the blood flows for a while in one direction, then stops, and flows in the opposite direction; just as through a rudely-organized society, the distribution of merchandise is slowly carried on by great fairs, occurring in different localities, to and from which the currents periodically set. Only animals of tolerably complete organizations, like advanced communities, are permeated by constant currents that are definitely directed. In living bodies, the local and variable currents disappear when there grow up great centres of circulation, generating more powerful currents, by a rhythm which ends in a quick, regular pulsation. And when in social bodies, there arise great centres of commercial activity, producing and exchanging large quantities of commodities, the rapid and continuous streams drawn in and emitted by these centres, subdue all minor and local circulations: the slow rhythm of fairs merges into the faster one of weekly markets, and in the chief centres of distribution, weekly markets merge into daily markets; while in place of the languid transfer from place to place, taking place at first weekly, then twice or thrice a week, we by-and-by get daily transfer, and finally transfer many times a day — the original sluggish, irregular rhythm, becomes a rapid, equable pulse.

Mark, too, that in both cases the increased activity, like the greater perfection of structure, is much less conspicuous at the periphery of the vascular system. On main lines of railway, we have, perhaps, a score trains in each direction daily, going at from thirty to fifty miles an hour; as, through the great arteries, the blood rushes rapidly in successive gushes. Along high roads, there move vehicles conveying men and commodities with much less, though still considerable, speed, and with a much less decided rhythm; as, in the smaller arteries, the speed of the blood is greatly diminished, and the pulse less conspicuous. In parish-roads, narrow, less complete, and more tortuous, the rate of movement is further decreased and the rhythm scarcely traceable; as in the ultimate arteries. In those still more imperfect by roads which lead from these parish-roads to scattered farmhouses and cottages, the motion is yet slower and very irregular; just as we find it in the capillaries. While along the field-roads, which, in their unformed, unfenced state, are typical of *lacunæ,* the movement is the slowest, the most irregular, and the most infrequent; as it is, not only in the primitive *lacunæ* of animals, and societies, but as it is also in those *lacunæ* in which the vascular system ends among extensive families of inferior creatures.

Thus, then, we find between the distributing systems of living bodies and the distributing systems of bodies politic, wonderfully close parallelisms. In the lowest forms of individual and social organisms, there exist neither prepared nutritive matters nor distributing appliances; and in both, these, arising as necessary accompaniments of the differentiation of parts, approach perfection as this differentiation approaches completeness. In animals, as in societies, the distributing agencies begin to show themselves at the same relative periods, and in the same relative positions. In the one, as in the other, the nutritive materials circulated, are at first crude and simple, gradually become better elaborated and more heterogeneous, and have eventually added to them a new element facilitating the nutritive processes. The channels of communication pass through similar phases of development, which bring them to analogous forms. And the directions, rhythms, and rates of circulation, progress by like steps to like final conditions.

We come at length to the nervous system. Having noticed the primary differentiation of societies into the governing and governed classes, and observed its analogy to the differentiation of the two primary tissues which respectively develope into organs of external action and organs of alimentation; having noticed some of the leading analogies between the development of industrial arrangements and that of the alimentary apparatus; and having, above, more fully traced the analogies between the distributing systems, social and individual; we have now to compare the appliances by which a society, as a whole, is regulated, with those by which the movements of an individual creature are regulated. We shall find here, parallelisms equally striking with those already detailed.

The class out of which governmental organization originates, is, as we have said, analogous in its relations to the ectoderm of the lowest animals and of embryonic forms. And as this primitive membrane, out of which the nervo-muscular system is evolved, must, even in the first stage of its differentiation, be slightly distinguished from the rest by that greater impressibility and contractility characterizing the organs to which it gives rise; so, in that superior class which is eventually transformed into the directo-executive system of a society (its legislative and defensive appliances), does there exist in the beginning, a larger endowment of the capacities required for these higher social functions. Always, in rude assemblages of men, the strongest, most courageous, and most sagacious, become rulers and leaders; and, in a tribe of some standing, this results in the establishment of a dominant class, characterized on the average by those mental and bodily qualities which fit them for deliberation and vigorous combined action. Thus that greater impressibility and contractility, which in the rudest animal types characterize the units of the ectoderm, characterize also the units of the primitive social ectoderm; since impressibility and contractility are the respective roots of intelligence and strength.

Again, in the unmodified ectoderm, as we see it in the *Hydra,* the units are all endowed both with impressibility and contractility; but as we ascend to higher types of organization, the ectoderm differentiates into classes of units which divide those two functions between them: some, becoming exclusively impressible, cease to be contractile; while some, becoming exclusively contractile, cease to be impressible. Similarly with societies. In an aboriginal tribe, the directive and executive functions are diffused in a mingled form throughout the whole governing class. Each minor chief commands those under him, and if need be, himself coerces them into obedience. The council of chiefs itself carries out on the battle-field its own decisions. The head chief not only makes laws, but administers justice with his own hands. In larger and more settled communities, however, the directive and executive agencies begin to grow distinct from each other. As fast as his duties accumulate, the head chief or king confines himself more and more to directing public affairs, and leaves the execution of his will to others: he deputes others to enforce submission, to inflict punishments, or to carry out minor acts of offence and defence; and only on occasions when, perhaps, the safety of the society and his own supremacy are at stake, does he begin to act as well as direct. As this differentiation establishes itself, the characteristics of the ruler begin to change. No longer, as in an aboriginal tribe, the strongest and most daring man, the tendency is for him to become the man of greatest cunning, foresight, and skill in the management of others; for in societies that have advanced beyond the first stage, it is chiefly such qualities that insure success in gaining supreme power, and holding it against internal and external enemies. Thus that member of the governing class who comes to be the chief directing agent, and so plays the same part that a rudimentary nervous centre does in an unfolding organism, is usually one endowed with some superiorities of nervous organization.

In those somewhat larger and more complex communities possessing, perhaps, a separate military class, a priesthood, and dispersed masses of population requiring local control, there necessarily grow up subordinate governing agents; who as their duties accumulate, severally become more directive and less executive in their characters. And when, as commonly happens, the king begins to collect round himself advisers who aid him by communicating information, preparing subjects for his judgment, and issuing his orders; we may say that the form of organization is comparable to one very general among inferior types of animals, in which there exists a chief ganglion with a few dispersed minor ganglia under its control.

The analogies between the evolution of governmental structures in societies, and the evolution of governmental structures in living bodies, are, however, more strikingly displayed during the formation of nations by the coalescence of small communities — a process already shown to be, in several respects, parallel to the development of those creatures that primarily consist of many like segments. Among other points of community between the successive rings which make up the body in the lower *Articulata, is* the possession of similar pairs of ganglia. These pairs of ganglia, though united together by nerves, are very incompletely dependent on any general controlling power. Hence it results that when the body is cut in two, the hinder part continues to move forward under the propulsion of its numerous legs; and that when the chain of ganglia has been divided without severing the body, the hind limbs may be seen trying to propel the body in one direction, while the fore limbs are trying to propel it in another. Among the higher *Articulata,* however, a number of the anterior pairs of ganglia, besides growing larger, unite in one mass; and this great cephalic ganglion, becoming the co-ordinator of all the creature's movements, there no longer exists much local in dependence.

Now may we not in the growth of a consolidated kingdom out of petty sovereignties or baronies, observe analogous changes? Like the chiefs and primitive rulers above described, feudal lords, exercising supreme power over their respective groups of retainers, discharge functions analogous to those of rudimentary nervous centres; and we know that at first they, like their analogues, are distinguished by superiorities of directive and executive organization. Among these local governing centres, there is, in early feudal times, very little subordination. They are in frequent antagonism; they are individually restrained chiefly by the influence of large parties in their own class; and are but imperfectly and irregularly subject to that most powerful member of their order who has gained the position of head suzerain or king. As the growth and organization of the society progresses, these local directive centres fall more and more under the control of a chief directive centre. Closer commercial union between the several segments, is accompanied by closer governmental union; and these minor rulers end in being little more than agents who administer, in their several localities, the laws made by the supreme ruler: just as the local ganglia above described, eventually become agents which enforce, in their respective segments, the orders of the cephalic ganglion.

The parallelism holds still further. We remarked above, when speaking of the rise of aboriginal kings, that in proportion as their territories and duties increase, they are obliged not only to perform their executive functions by deputy, but also to gather round themselves advisers to aid them in their directive functions; and that thus, in place of a solitary governing unit, there grows up a group of governing units, comparable to a ganglion consisting of many cells. Let us here add, that the advisers and chief officers who thus form the rudiment of a ministry, tend from the beginning to exercise a certain control over the ruler. By the information they give and the opinions they express, they sway his judgment and affect his commands. To this extent he therefore becomes a channel through which are communicated the directions originating with them; and in course of time, when the advice of ministers becomes the acknowledged source of his actions, the king assumes very much the character of an automatic centre, reflecting the impressions made on him from without.

Beyond this complication of governmental structure, many societies do not progress; but in some, a further development takes place. Our own case best illustrates this further development, and its further analogies. To kings and their ministries have been added, in England, other great directive centres, exercising a control which, at first small, has been gradually becoming predominant: as with the great governing ganglia that especially distinguish the highest classes of living beings. Strange as the assertion will be thought, our Houses of Parliament discharge in the social economy, functions that are in sundry respects comparable to those discharged by the cerebral masses in a vertebrate animal. As it is in the nature of a single ganglion to be affected only by special stimuli from particular parts of the body; so it is in the nature of a single ruler to be swayed in his acts by exclusive personal or class interests. As it is in the nature of an aggregation of ganglia, connected with the primary one, to convey to it a greater variety of influences from more numerous organs, and thus to make its acts conform to more numerous requirements; so it is in the nature of a king surrounded by subsidiary controlling powers, to adapt his rule to a greater number of public exigencies. And as it is in the nature of those great and latest-developed ganglia which distinguish the higher animals, to interpret and combine the multiplied and varied impressions conveyed to them from all parts of the system, and to regulate the actions in such way as duly to regard them all; so it is in the nature of those great and latest-developed legislative bodies which distinguish the most advanced societies, to interpret and combine the wishes and complaints of all classes and localities, and to regulate public affairs as much as possible in harmony with the general wants.

The cerebrum co-ordinates the countless heterogeneous considerations which affect the present and future welfare of the individual as a whole; and the legislature co-ordinates the countless heterogeneous considerations which affect the immediate and remote welfare of the whole community. We may describe the office of the brain as that of *averaging* the interests of life, physical, intellectual, moral, social; and a good brain is one in which the desires answering to these respective interests are so balanced, that the conduct they jointly dictate, sacrifices none of them. Similarly, we may describe the office of a Parliament as that of *averaging* the interests of the various classes in a community; and a good Parliament is one in which the parties answering to these respective interests are so balanced, that their united legislation concedes to each class as much as consists with the claims of the rest. Besides being comparable in their duties, these great directive centres, social and individual, are comparable in the processes by which their duties are discharged.

It is now an acknowledged truth in psychology, that the cerebrum is not occupied with direct impressions from without, but with the ideas of such impressions: instead of the actual sensations produced in the body, and directly appreciated by the sensory ganglia or primitive nervous centres, the cerebrum receives only the representations of these sensations; and its consciousness is called *representative* consciousness, to distinguish it from the original or *presentative* consciousness. Is it not significant that we have hit on the same word to distinguish the function of our House of Commons? We call it a *representative* body, because the interests with which it deals — the pains and pleasures about which it consults — are not directly presented to it, but represented to it by its various members; and a debate is a conflict of representations of the evils or benefits likely to follow from a proposed course — a description which applies with equal truth to a debate in the individual consciousness. In both cases, too, these great governing masses take no part in the executive functions. As, after a conflict in the cerebrum, those desires which finally predominate, act on the subjacent ganglia, and through their instrumentality determine the bodily actions; so the parties which, after a parliamentary struggle, gain the victory, do not themselves carry out their wishes, but get them carried out by the executive divisions of the Government. The fulfilment of all legislative decisions still devolves on the original directive centres — the impulse passing from the Parliament to the Ministers, and from the Ministers to the King, in whose name everything is done; just as those smaller, first-developed ganglia, which in the lowest vertebrata are the chief controlling agents, are still, in the brains of the higher vertebrata, the agents through which the dictates of the cerebrum are worked out.

Moreover, in both cases these original centres become increasingly automatic. In the developed vertebrate animal, they have little function beyond that of conveying impressions to, and executing the determinations of, the larger centres. In our highly organized government, the monarch has long been lapsing into a passive agent of Parliament; and now, ministers are rapidly falling into the same position.

Nay, between the two cases there is a parallelism, even in respect of the exceptions to this automatic action. For in the individual creature, it happens that under circumstances of sudden alarm, as from a loud sound close at hand, an unexpected object starting up in front, or a slip from insecure footing, the danger is guarded against by some quick involuntary jump, or adjustment of the limbs, that takes place before there is time to consider the impending evil, and take deliberate measures to avoid it: the rationale of which is, that these violent impressions produced on the senses, are reflected from the sensory ganglia to the spinal cord and muscles, without, as in ordinary cases, first passing through the cerebrum. In like manner, on national emergencies, calling for prompt action, the King and Ministry, not having time to lay the matter before the great deliberative bodies, themselves issue commands for the requisite movements or precautions: the primitive, and now almost automatic, directive centres, resume for a moment their original uncontrolled power. And then, strangest of all, observe that in either case there is an afterprocess of approval or disapproval. The individual on recovering from his automatic start, at once contemplates the cause of his fright; and, according to the case, concludes that it was well he moved as he did, or condemns himself for his groundless alarm. In like manner, the deliberative powers of the State, discuss, as soon as may be, the unauthorized acts of the executive powers and, deciding that the reasons were or were not sufficient, grant or withhold a bill of indemnity.[22]

---

[22] It may be well to warn the reader against an error fallen into by one who criticised this essay on its first publication — the error of supposing that the analogy here intended to be drawn, is a specific analogy between the organization of society in England, and the human organization. As said at the outset, no-such specific analogy exists. The above

Thus far in comparing the governmental organization of the body politic with that of an individual body, we have considered only the respective co-ordinating centres. We have yet to consider the channels through which these co-ordinating centres receive information and convey commands. In the simplest societies, as in the simplest organisms, there is no "internuncial apparatus," as Hunter styled the nervous system. Consequently, impressions can be but slowly propagated from unit to unit throughout the whole mass. The same progress, however, which, in animal-organization, shows itself in the establishment of ganglia or directive centres, shows itself also in the establishment of nerve-threads, through which the ganglia receive and convey impressions, and so control remote organs. And in societies the like eventually takes place.

After a long period during which the directive centres communicate with various parts of the society through other means, there at last comes into existence an "internuncial apparatus," analogous to that found in individual bodies. The comparison of telegraph-wires to nerves, is familiar to all. It applies, however, to an extent not commonly supposed. We do not refer to the near alliance between the subtle forces employed in the two cases; though it is now held that the nerve-force, if not literally electric, is still a special form of electric action, related to the ordinary form much as magnetism is. But we refer to the structural arrangements of our telegraph-system. Thus, throughout the vertebrate sub-kingdom, the great nerve-bundles diverge from the vertebrate axis, side by side will the great arteries; and similarly, our groups of telegraph-wires are carried along the sides of our railways. The most striking parallelism, however, remains. Into each great bundle of nerves, as it leaves the axis of the body along with an artery, there enters a branch of the sympathetic nerve; which branch, accompanying the artery throughout its ramifications, has the function of regulating its diameter and otherwise controlling the flow of blood through it according to the

---

parallel, is one between the most-developed systems of governmental organization, individual and social; and the vertebrate type is instanced, merely as exhibiting this most-developed system. If any specific comparison were made, which it cannot rationally be, it would be to some much lower vertebrate form than the human.

local requirements. Analogously, in the group of telegraph-wires running alongside each railway, there is one for the purpose of regulating the traffic — for retarding or expediting the flow of passengers and commodities, as the local conditions demand. Probably, when our now rudimentary telegraph-system is fully developed, other analogies will be traceable.

Such, then, is a general outline of the evidence which justifies, in detail, the comparison of societies to living organisms. That they gradually increase in mass; that they become little by little more complex; that at the same time their parts grow more mutually dependent; and that they continue to live and grow as wholes, while successive generations of their units appear and disappear; are broad peculiarities which bodies politic display, in common with all living bodies; and in which they and living bodies differ from everything else. And on carrying out the comparison in detail, we find that these major analogies involve many minor analogies, far closer than might have been expected. To these we would gladly have added others. We had hoped to say something respecting the different types of social organization, and something also on social metamorphoses; but we have reached our assigned limits.

XI
USE AND BEAUTY

In one of his essays, Emerson remarks, that what Nature at one time provides for use, she afterwards turns to ornament; and he cites in illustration the structure of a sea-shell, in which the parts that have for a while formed the mouth are at the next season of growth left behind, and become decorative nodes and spines.

It has often occurred to me that this same remark might be extended to the progress of Humanity. Here, too, the appliances of one era serve as embellishments to the next. Equally in institutions, creeds, customs, and superstitions, we may trace this evolution of beauty out of what was once purely utilitarian.

The contrast between the feeling with which we regard portions of the Earth's surface still left in their original state, and the feeling with which the savage regarded them, is an instance that naturally comes first in order of time. If any one walking over Hampstead Heath, will note how strongly its picturesqueness is brought out by contrast with the surrounding cultivated fields and the masses of houses lying in the distance; and will further reflect that, had this irregular gorse-covered surface extended on all aides to the horizon, it would have looked dreary and prosaic rather than pleasing; he will see that to the primitive man a country so clothed presented no beauty at all. To him it was merely a haunt of wild animals, and a ground out of which roots might be dug. What have become for us places of relaxation and enjoyment — places for afternoon strolls and for gathering flowers — were his places for labour and food, probably arousing in his mind none but utilitarian associations.

Ruined castles afford an obvious instance of this metamorphosis of the useful into the beautiful. To feudal barons and their retainers, security was the chief, if not the only end, sought in choosing the sites and styles of their strongholds. Probably they aimed as little at the picturesque as do the builders of cheap brick houses in our modern towns. Yet what where erected for shelter and safety, and what in those early days fulfilled an important function in the social economy, have now assumed a purely ornamental character. They serve as scenes for picnics; pictures of them decorate our drawing-rooms; and each supplies its surrounding districts with legends for Christmas Eve.

Following out the train of thought suggested by this last illustration, we may see that not only do the material exuviæ of past social states become the ornaments of our landscapes; but that past habits, manners, and arrangements, serve as ornamental elements in our literature. The tyrannies that, to the serfs who bore them, were harsh and dreary facts; the feuds which, to those who took part in them, were very practical life-and-death affairs; the mailed, moated, sentinelled security that was irksome to the nobles who needed it; the imprisonments, and tortures, and escapes, which were stern and quite prosaic realities to all concerned in them; have become to us material for romantic tales — material which when woven into Ivanhoes and Marmions, serves for amusement in leisure hours, and become poetical by contrast with our daily lives.

Thus, also, is it with extinct creeds. Stonehenge, which in the hands of the Druids had a governmental influence over men, is in our day a place for antiquarian excursions; and its attendant priests are worked up into an opera. Greek sculptures, preserved for their beauty in our galleries of art, and copied for the decoration of pleasure grounds and entrance halls, once lived in men's minds as gods demanding obedience; as did also the grotesque idols that now amuse the visitors to our museums.

Equally marked is this change of function in the case of minor superstitions. The fairy lore, which in past times was matter of grave belief, and held sway over people's conduct, has since been transformed into ornament for *A Midsummer Night's Dream, The Tempest, The Fairy Queen,* and endless small tales and poems; and still affords subjects for children's story-books, themes for ballets, and plots for Planché's burlesques. Gnomes, and genii, and afrits, losing all their terrors, give piquancy to the woodcuts in our illustrated edition of the *Arabian Nights.* While ghost-stories, and tales of magic and witchcraft, after serving to amuse boys and girls in their leisure hours, become matter for jocose allusions that enliven tea-table conversation.

Even our serious literature and our speeches are very generally relieved by ornaments drawn from such sources. A Greek myth is often used as a parallel by which to vary the monotony of some grave argument. The lecturer breaks the dead level of his practical discourse by illustrations drawn from bygone customs, events, or beliefs. And metaphors, similarly derived, give brilliancy to political orations, and to *Times* leading articles.

Indeed, on careful inquiry, I think it will be found that we turn to purposes of beauty most bygone phenomena that are at all conspicuous. The busts of great men in our libraries, and their tombs in our churches; the once useful but now purely ornamental heraldic symbols; the monks, nuns, and convents, that give interest to a certain class of novels; the bronze mediæval soldiers used for embellishing drawing-rooms; the gilt Apollos that recline on timepieces; the narratives that serve as plots for our great dramas; and the events that afford subjects for historical pictures; — these and such like illustrations of the metamorphosis of the useful into the beautiful, are so numerous as to suggest that, did we search diligently enough, we should find that in some place, or under some circumstances, nearly every notable product of the past has assumed a decorative character.

And here the mention of historical pictures reminds me that an inference may be drawn from all this, bearing directly on the practice of art. It has of late years been a frequent criticism upon our historical painters, that they err in choosing their subjects from the past; and that, would they found a genuine and vital school, they must render on canvas the life and deeds and aims of our own time. If, however, there be any significance in the foregoing facts, it seems doubtful whether this criticism is a just one. For if it be the process of things, that what has performed some practical function in society during one era, becomes available for ornament in a subsequent one; it almost follows that, conversely, whatever is performing some practical function now, or has very recently performed one, does not possess the ornamental character; and is, consequently, inapplicable to any purpose of which beauty is the aim, or of which it is a needful ingredient.

Still more reasonable will this conclusion appear, when we consider the nature of this process by which the useful is changed into the ornamental. An essential pre-requisite to all beauty is *contrast.* To obtain artistic effect, light must be put in juxtaposition with shade, bright colours with dull colours, a fretted surface with a plain one. *Forte* passages in music must have *piano* passages to relieve them; concerted pieces need interspersing with solos; and rich chords must not be continuously repeated. In the drama we demand contrast of characters, of scenes, of sentiment, of style. In prose composition an eloquent passage should have a comparatively plain setting; and in poems great effect is obtained by occasional change of versification. This general principle will, I think, explain the transformation of the bygone useful into the present beautiful. It is by virtue of their contrast with our present modes of life, that past modes of life look interesting and romantic. Just as a picnic, which is a temporary return to an aboriginal condition, derives, from its unfamiliarity, a certain poetry which it would not have were it habitual; so, everything ancient gains, from its relative novelty to us, an element of interest. Gradually as, by the growth of society, we leave behind the customs, manners, arrangements, and all the products, material and mental, of a bygone age — gradually as we recede from these so far that there arises a conspicuous difference between them and those we are familiar with; so gradually do they begin to assume to us a poetical aspect, and become applicable for ornament. And hence it follows that things and events which are close to us, and which are accompanied by associations of ideas not markedly contrasted with our ordinary associations we relatively inappropriate for purposes of art.

XII
THE SOURCES OF ARCHITECTURAL TYPES

When lately looking through the gallery of the Old Water-Colour Society, I was struck with the incongruity produced by putting regular architecture into irregular scenery. In one case, where the artist had introduced a perfectly symmetrical Grecian edifice into a mountainous and somewhat wild landscape, the discordant effect was particularly marked. "How very unpicturesque," said a lady to her friend, as they passed; showing that I was not alone in my opinion. Her phrase, however, set me speculating. Why unpicturesque? Picturesque means, like a picture — like what men choose for pictures. Why then should this be not fit for a picture?

Thinking the matter over, it seemed to me that the artist had sinned against that unity which is essential to a good picture. When the other constituents of a landscape have irregular forms, any artificial structure introduced must have an irregular form, that it may seem *part* of the landscape. The same general character must pervade it and surrounding objects; otherwise it, and the scene amid which it stands, become not *one* thing but *two* things; and we say it looks out of place. Or, speaking psychologically, the associated ideas called up by a building with its wings, windows, and all its parts symmetrically disposed, differ widely from the ideas associated with an entirely irregular landscape; and the one set of ideas tends to banish the other.

Pursuing the train of thought, sundry illustrative facts came to my mind. I remembered that a castle, which is more irregular in outline than any other kind of building, pleases us most when seated amid crags and precipices; while a castle on a plain seems an incongruity. The partly regular and partly-irregular forms of our old farm-houses, and our gabled gothic manors and abbeys, appear quite in harmony with an undulating, wooded country. In towns we prefer symmetrical architecture; and in towns it produces in us no feeling of incongruity, because all surrounding things — men, horses, vehicles — are symmetrical also.

And here I was reminded of a notion that has frequently recurred to me; namely, that there is some relationship between the several kinds of architecture and the several classes of natural objects. Buildings in the Greek and Roman styles seem, in virtue of their symmetry, to take their type from animal life. In the partly-irregular Gothic, ideas derived from the vegetable world appear to predominate. And wholly irregular buildings, such as castles, may be considered as having inorganic forms for their basis.

Whimsical as this speculation looks at first sight, it is countenanced by numerous facts. The connexion between symmetrical architecture and animal forms, may be inferred from the *kind* of symmetry we expect, and are satisfied with, in regular buildings. Thus in a Greek temple we require that the front shall be symmetrical in itself, and that the two flanks shall be alike; but we do not look for uniformity between the flanks and the front, nor between the front and the back. The identity of this symmetry with that found in animals is obvious. Again, why is it that a building making any pretension to symmetry displeases us if not quite symmetrical? Probably the reply will be — Because we see that the designer's idea is not fully carried out; and that hence our love of completeness is offended. But then there come the further questions — How do we know that the architect's conception was symmetrical? Whence comes this notion of symmetry which we have, and which we attribute to him? Unless we fall back upon the old doctrine of innate ideas, we must admit that the idea of bilateral symmetry is derived from without; and to admit this is to admit that it is derived from the higher animals.

That there is some relationship between Gothic architecture and vegetable forms is a position generally admitted. The often-remarked analogy between a groined nave and an avenue of trees with interlacing branches, shows that the fact has forced itself on men's observation. It is not only in this analogy, however, that the kinship is seen. It is seen still better in the essential characteristic of Gothic; namely, what is termed its *aspiring* tendency. That predominance of vertical lines which so strongly distinguishes Gothic from other styles, is the most marked peculiarity of trees, when compared with animals or rocks. To persons of active imagination, a tall Gothic tower, with its elongated apertures and clusters of thin projections running from bottom to top, suggests a vague notion of growth.

Of the alleged connexion between inorganic forms and the wholly irregular and the castellated styles of building, we have, I think, some proof in the fact that when an edifice is irregular, the *more* irregular it is the more it pleases us. I see no way of accounting for this fact, save by supposing that the greater the irregularity the more strongly are we reminded of the inorganic forms typified, and the more vividly are aroused the agreeable ideas of rugged and romantic scenery associated with those forms.

Further evidence of these several relationships of styles of architecture to classes of natural objects, is supplied by the kinds of decoration they respectively represent. The public buildings of Greece, while characterized in their outlines by the bilateral symmetry seen in the higher animals, have their pediments and entablatures covered with sculptured men and beasts. Egyptian temples and Assyrian palaces, while similarly symmetrical in their general plan, are similarly ornamented on their walls and at their doors. In Gothic, again, with its grove-like ranges of clustered columns, we find rich foliated ornaments abundantly employed. And accompanying the totally irregular, inorganic outlines of old castles, we see neither vegetable nor animal decorations. The bare, rock-like walls are surmounted by battlements, consisting of almost plain blocks, which remind us of the projections on the edge of a rugged cliff.

But perhaps the most significant fact is the harmony that may be observed between each type of architecture and the scenes in which it is indigenous. For what is the explanation of this harmony, unless it be that the predominant character of surrounding things has, in some way, determined the mode of building adopted?

That the harmony exists is clear. Equally in the cases of Egypt, Assyria, Greece, and Rome, town life preceded the construction of the symmetrical buildings that have come down to us. And town life is one in which, as already observed, the majority of familiar objects are symmetrical. We instinctively feel the naturalness of this association. Out amid the fields, a formal house, with a central door flanked by an equal number of windows to right and left, strikes us as unrural — looks as though transplanted from a street; and we cannot look at one of those stuccoed villas, with mock windows carefully arranged to balance the real ones, without being reminded of the suburban residence of a retired tradesman.

In styles indigenous in the country, we not only find the general irregularity characteristic of surrounding things, but we may trace some kinship between each kind of irregularity and the local circumstances. We see the broken rocky masses amid which castles are commonly placed, mirrored in their stern, inorganic forms. In abbeys, and suchlike buildings, which are commonly found in comparatively sheltered districts, we find no such violent dislocations of masses and outlines; and the nakedness appropriate to the fortress is replaced by decorations reflecting the neighbouring woods. Between a Swiss cottage and a Swiss view there is an evident relationship. The angular roof, so bold and so disproportionately large when compared to other roofs, reminds one of the adjacent mountain peaks; and the broad overhanging eaves have a sweep and inclination like those of the lower branches of a pine tree. Consider, too, the apparent kinship between the flat roofs that prevail in Eastern cities, interspersed with occasional minarets, and the plains that commonly surround them, dotted here and there by palm trees. You cannot contemplate a picture of one of these places, without being struck by the predominance of horizontal lines, and their harmony with the wide stretch of the landscape.

That the congruity here pointed out should hold in every case must not be expected. The Pyramids, forex ample, do not seem to come under this generalization. Their repeated horizontal lines do indeed conform to the flatness of the neighbouring desert; but their general contour seems to have no adjacent analogue. Considering, however, that migrating races, carrying their architectural systems with them, would naturally produce buildings having no relationship to their new localities; and that it is not always possible to distinguish styles which are indigenous, from those which are naturalized; numerous anomalies must be looked for.

The general idea above illustrated will perhaps be somewhat misinterpreted. Possibly some will take the proposition to be that men *intentionally* gave to their buildings the leading characteristics of neighbouring objects. But this is not what is meant. I do not suppose that they did so in times past, any more than they do so now. The hypothesis is, that in their choice of forms men are unconsciously influenced by the forms encircling them. That flat-roofed, symmetrical architecture should have originated in the East, among pastoral tribes surrounded by their herds and by wide plains, seems to imply that the builders were swayed by the horizontality and symmetry to which they were habituated. And the harmony which we have found to exist in other cases between indigenous styles and their localities, implies the general action of like influences. Indeed, on considering the matter psychologically, I do not see how it could well be otherwise. For as all conceptions must be made up of images, and parts of images, received through the senses — as it is impossible for a man to conceive any design save one of which the elements have come into his mind from without; and as his imagination will most readily run in the direction of his habitual perceptions; it follows, almost necessarily, that the characteristic which predominates in these habitual perceptions must impress itself on his design.

XIII
THE USE OF ANTHROPOMORPHISM

That long fit of indignation which seizes all generous natures when in youth they begin contemplating human affairs, having fairly spent itself, there slowly grows up a perception that the institutions, beliefs, and forms so vehemently condemned are not wholly bad. This reaction runs to various lengths. In some, merely to a comparative contentment with the arrangements under which they live. In others to a recognition of the fitness that exists between each people and its government, tyrannical as that may be. In some, again, to the conviction, that hateful though it is to us, and injurious as it would be now, slavery was once beneficial — was one of the necessary phases of human progress. Again, in others, to the suspicion that great benefit has indirectly arisen from the perpetual warfare of past times; insuring as this did the spread of the strongest races, and so providing good raw material for civilization. And in a few this reaction ends in the generalization that all modes of human thought and action subserve, in the times and places in which they occur, some useful function: that though bad in the abstract, they are relatively good — are the best which the then existing conditions admit of.

A startling conclusion to which this faith in the essential beneficence of things commits us, is that the religions creeds through which mankind successively pass, are, during the eras in which they are severally held, the best that could be held; and that this is true, not only of the latest and most refined creeds, but of all, even to the earliest and most gross. Those who regard men's faiths as given to them from without — as having origins either directly divine or diabolical, and who, considering their own as the sole example of the one, class all the rest under the other, will think this a very shocking opinion. I can imagine, too, that many of those who have abandoned current theologies, and now regard religions as so many natural products of human nature — men who, having lost that antagonism towards their old creed which they felt while shaking themselves free from it, can now see that it was highly beneficial to past generations, and is beneficial still to a large part of mankind; — I can imagine even these hardly prepared to admit that all religions, down to the lowest Fetichism, have, in their places, fulfilled useful functions. If such, however, will consistently develop their ideas, they will find this inference involved.

For if it be true that humanity in its corporate as well as in its individual aspect, is a growth and not a manufacture, it is obvious that during each phase men's theologies, as well as their political and social arrangements, must be determined into such forms as the conditions require. In the one case as in the other, by a tentative process, things from time to time re-settle themselves in a way that best consists with national equilibrium. As out of plots and the struggles of chieftains, it continually results that the strongest gets to the top, and by virtue of his proved superiority ensures a period of quiet, and gives society time to grow; as out of incidental expedients there periodically arise new divisions of labour, which get permanently established only by serving men's wants better than the previous arrangements did; so, the creed which each period evolves is one more in conformity with the needs of the time than the creed which preceded it. Not to rest in general statements, however, let us consider why this must be so. Let us see whether, in the genesis of men's ideas of deity, there is not involved a necessity to conceive of deity under the aspect most influential with them.

It is now generally admitted that a more or less idealized humanity is the form which every conception of a personal God must take. Anthropomorphism is an inevitable result of the laws of thought. We cannot take a step towards constructing an idea of God without the ascription of human attributes. We cannot even speak of a divine will without assimilating the divine nature to our own; for we know nothing of volition save as a property of our own minds.

While this anthropomorphic tendency, or rather necessity, is manifested by themselves with sufficient grossness — a grossness that is offensive to those more advanced — Christians are indignant at the still grosser manifestations of it seen among uncivilized men. Certainly, such conceptions as those of some Polynesians, who believe that their gods feed on the souls of the dead, or as those of the Greeks, who ascribed to the personages of their Pantheon every vice, from domestic cannibalism downward, are repulsive enough. But if, ceasing to regard these notions from the outside, we more philosophically regard them from the inside — if we consider how they looked to believers, and observe the relationships they bore to the natures and needs of such; we shall begin to think of them with some tolerance. The question to be answered is, whether these beliefs were beneficent in their effects on those who held them; not whether they would be beneficent for us, or for perfect men; and to this question the answer must be that while absolutely bad, they were relatively good.

For is it not obvious that the savage man will be most effectually controlled by his fears of a savage deity? Must it not happen, that if his nature requires great restraint, the supposed consequences of transgression, to be a check upon him, must be proportionately terrible; and for these to be proportionately terrible, must not his god be conceived as proportionately cruel and revengeful? Is it not well that the treacherous, thievish, lying Hindoo should believe in a hell where the wicked are boiled in cauldrons, rolled down mountains bristling with knives, and sawn asunder between flaming iron posts? And that there may be provided such a hell, is it not needful that he should believe in a divinity delighting in human immolations and the self-torture of fakirs? Does it not seem clear that during the earlier ages in Christendom, when men's feelings were so hard that a holy father could describe one of the delights of heaven to be the contemplation of the torments of the damned — does it not seem clear that while the general nature was so unsympathetic, there needed, to keep men in order, all the prospective tortures described by Dante, and a deity implacable enough to inflict them?

And if, as we thus see, it is well for the savage man to believe in a savage god, then we may also see the great usefulness of this anthropomorphic tendency; or, as before said, necessity. We have in it another illustration of that essential beneficence of things visible everywhere throughout nature. From this inability under which we labour to conceive of a deity save as some idealization of ourselves, it inevitably results that in each age, among each people, and to a great extent in each individual, there must arise just that conception of deity best adapted to the needs of the case. If, being violent and bloodthirsty, the nature be one calling for stringent control, it evolves the idea of a ruler still more violent and bloodthirsty, and fitted to afford this control. When, by ages of social discipline, the nature has been partially humanized, and the degree of restraint required has become less, the diabolical characteristics before ascribed to the deity cease to be so predominant in the conception of him. And gradually, as all need for restraint disappears, this conception approximates towards that of a purely beneficent necessity. Thus, man's constitution is in this, as in other respects, self-adjusting, self-balancing. The mind itself evolves a compensating check to its own movements; varying always in proportion to the requirement. Its centrifugal and its centripetal forces are necessarily in correspondence, because the one generates the other. And so we find that the forms of both religious and secular rule follow the same law. As an ill-controlled national character produces a despotic terrestrial government, so also does it produce a despotic celestial government — the one acting through the senses, the other through the imagination; and in the converse case the same relationship holds good.

Organic as this relationship is in its origin, no artificial interference can permanently affect it. Whatever perturbations an external agency may seem to produce, they are soon neutralized in fact, if not in appearance. I was recently struck with this in reading a missionary account of the "gracious visitations of the Holy Spirit at Vewa," one of the Feejee islands. Describing a "penitent meeting," the account says: —

"Certainly the feelings of the Vewa people were not ordinary. They literally roared for hours together for the disquietude of their souls. This frequently terminated in fainting from exhaustion, which was the only respite some of them had till they found peace. They no sooner recovered their consciousness than they prayed themselves first into an agony, then again into a state of entire insensibility."

Now these Feejee islanders are the most savage of all the uncivilized races. They are given to cannibalism, in fanticide, and human sacrifices; they are so bloodthirsty and so treacherous, that members of the same family dare not trust each other; and, in harmony with these characteristics, they have for their aboriginal god, a serpent. Is it not clear then, that these violent emotions which the missionaries describe, these terrors and agonies of despair which they rejoiced over, were nothing but the worship of the old god under a new name? Is it not clear that these Feejees had simply understood those parts of the Christian creed which agree in spirit with their own — the vengeance, the perpetual torments, the diabolism of it; that these, harmonizing with their natural conceptions of divine rule, were realized by them with extreme vividness; and that the extremity of the fear which made them "literally roar for hours together," arose from the fact that while they could fully take in and believe the punitive element, the merciful one was beyond their comprehension? This is the obvious inference. And it carries with it the further one, that in essence their new belief was merely their old one under a new form — the same substantial conception with a different history and different names.

However great, therefore, may be the seeming change adventitiously produced in a people's religion, the anthropomorphic tendency prevents it from being other than a superficial change — insures such modifications of the new religion as to give it all the potency of the old one — obscures whatever higher elements there may be in it until the people have reached the capability of being acted upon by them: and so, re-establishes the equilibrium between the impulses and the control they need. If any one requires detailed illustrations of this, he will find them in abundance in the history of the modifications of Christianity throughout Europe.

Ceasing then to regard heathen theologies from the personal point of view, and considering them solely with reference to the function they fulfil where they are indigenous, we must recognise them in common with all theologies, as good for their time and places; and this mental necessity which disables us from conceiving a deity save as some idealization of ourselves, we must recognise as the agency by which harmony is produced and maintained between every phase of human character and its religious creed.

Made in the USA
Monee, IL
04 November 2026

67391569R00203